What Every Engineer Should Know About Project Management

WHAT EVERY ENGINEER SHOULD KNOW

A Series

Editor

William H. Middendorf

Department of Electrical and Computer Engineering
University of Cincinnati
Cincinnati, Ohio

Vol. 1 What Every Engineer Should Know About Patents, *William G. Konold, Bruce Tittel, Donald F. Frei, and David S. Stallard*

Vol. 2 What Every Engineer Should Know About Product Liability, *James F. Thorpe and William H. Middendorf*

Vol. 3 What Every Engineer Should Know About Microcomputers: Hardware/ Software Design: A Step-by-Step Example, *William S. Bennett and Carl F. Evert, Jr.*

Vol. 4 What Every Engineer Should Know About Economic Decision Analysis, *Dean S. Shupe*

Vol. 5 What Every Engineer Should Know About Human Resources Management, *Desmond D. Martin and Richard L. Shell*

Vol. 6 What Every Engineer Should Know About Manufacturing Cost Estimating, *Eric M. Malstrom*

Vol. 7 What Every Engineer Should Know About Inventing, *William H. Middendorf*

Vol. 8 What Every Engineer Should Know About Technology Transfer and Innovation, *Louis N. Mogavero and Robert S. Shane*

Vol. 9 What Every Engineer Should Know About Project Management, *Arnold M. Ruskin and W. Eugene Estes*

Other volumes in preparation

What Every Engineer Should Know About Project Management

Arnold M. Ruskin
Partner
Claremont Consulting Group
Claremont, California

W. Eugene Estes
Director of Project Management
Dames and Moore
Los Angeles, California

MARCEL DEKKER, INC. New York and Basel

Library of Congress Cataloging in Publication Data

Ruskin, Arnold M.
What every engineer should know about project management.

(What every engineer should know; v. 9)
Bibliography: p. 151
Includes index.
1. Engineering--Management. I. Estes, W. Eugene, 1923- . II. Title. III. Series.
TA190.R87 1982 658.4'04 82-13089
ISBN 0-8247-1718-X

MARCEL DEKKER, INC.
270 Madison Avenue, New York, New York 10016

Current printing (last digit):
10

PRINTED IN THE UNITED STATES OF AMERICA

Preface

This book presents basic concepts and tools of projects and project management. Projects date from the earliest days of civilization and include the building of the Egyptian pyramids and the Roman roads and aqueducts. Today, projects are organized not only for building great public works but also for such diverse tasks as performing applied research, developing software, installing equipment, building extensive and complex systems, shutting down and renovating major facilities, and preparing proposals and studies. Few, if any, engineers can totally escape project work in the 1980s.

Our goals in this book are to provide a fundamental and broad view of project activity that applies to all engineers and to introduce some useful tools to the novice project manager. We have organized the book around the duties of the project manager because project management is what makes project work different from other technical work. The topics are important, however, to all who are associated with projects, whether they are project managers, supervisors of project managers, managers of subprojects and tasks, project staff members, others who support projects, or project customers. Likewise, the topics apply to engineers in all kinds of settings, such as industrial firms, governmental agencies, consulting firms, and literally all organizations where particular objectives must be accomplished and accomplished on time and within budget.

For simplicity, we refer to the person or group who is served by a project as the customer. The customer may in fact be a client of the project manager's organization, but it may just as likely be the project manager's boss, the organization's chief engineer, its marketing vice president, or some other individual or group within the organization. Whatever the case, the same principles apply.

All successful project work rests on a foundation of setting objectives, planning, directing and coordinating, controlling, reporting, and negotiating. We discuss all of these. Certain aspects of these topics are not limited to project work and might be considered by some to lie outside the realm of a book on project management. We include them because we want the reader to know how broad a project manager's duties are. People fall down in project management when they do not address their full range of duties and their consequences.

There is little that is truly mysterious about managing projects. It can only seem so because of the many facets involved. Our aim is to identify these facets and make them clear so that any mystery is removed. We hope that our reader will obtain the awareness, understanding, and basic equipment needed first to approach a project assignment with a healthy mix of respect and confidence and then to complete it successfully.

Arnold M. Ruskin
W. Eugene Estes

Table of Contents

List of Figures

List of Tables

About the Authors

ARNOLD M. RUSKIN, Ph.D., P.E., is co-founder and partner in Claremont Consulting Group, Claremont, California. He is also Adjunct Professor of Engineering and Program Coordinator of the Engineering Executive Program, UCLA, and Manager of Network Strategy Development, Jet Propulsion Laboratory, Pasadena. Dr. Ruskin is Book Review Editor for *Engineering Management International* and is the author of over 25 papers on a variety of topics in engineering and engineering management. Dr. Ruskin was the first chairman of the Engineering Management Committee of the American Society for Engineering Education and is a member of the American Institute of Chemical Engineers; the American Institute of Mining, Metallurgical, and Petroleum Engineers; the Project Management Institute; and the American Society for Engineering Management.

W. EUGENE ESTES, M.S., P.E., is a partner in Dames and Moore, Los Angeles, where he is Director of Project Management. His professional experience spans 35 years and has largely concerned the direct management of projects, the supervision of project managers, and the auditing of projects. Projects performed under his supervision have included research and development, design, construction, and major maintenance. He is a Fellow of the American Society of Civil Engineers and a member of the Project Management Institute and the Society of American Military Engineers.

What Every Engineer Should Know About Project Management

Introduction

Project management can be learned. While some people seem to know intuitively how to manage projects, most effective project managers learn their skills. This book explains how to manage projects so that their desired results are obtained on schedule and within budget.

Skillful project management involves knowing what is to be done, who is to do it, and when and how it should be done. This book considers all these factors and explains why as well. After all, there is little in the affairs of men and women that does not require judgement. If project managers are to apply their judgement wisely, they need to know the rationale for each task as well as the task itself.

The skills described in this book apply to projects large and small, to design efforts and construction jobs, to field studies, laboratory investigations, and software development. Indeed they apply to every kind of project. True, large or complicated projects offer more opportunities to overlook critical aspects than small or simple projects. But small and simple projects may be so stringently scheduled or budgeted that they provide no margin for error. In either case, therefore, project managers must pay attention to the factors discussed in the following chapters, whether explicitly or implicitly, if they are to ensure successful projects.

As a text on project management in general, we make no attempt to address the nuances peculiar to selected types of projects. Rather, we assume that our readers will supply their own insights into their particular situations and thus season our basic principles to suit their own tastes.

As noted above, this book is a mixture of the what, who, when, how, and why of project management. Some background, however, is given first so that the context of the what, who, etc., is clear. For this reason, the next chapter describes and characterizes projects. We then move in the following chapter, Chapter 2, to the roles and responsibilities of the project manager; in this chapter we begin to introduce some tools that the project manager can use and outline the functions of others that are substantial enough to warrant their own chapters. Beginning in Chapter 3 and continuing to Chapter 7, we take up individual tools and techniques that most project managers find occasion to use from time to time. Finally, in the epilogue we try to show the inter-dependence of the tools and techniques when they are used to manage projects.

Everyone manages projects, but some manage better than others. In many cases, the consequences of inept management are not terribly significant. If it takes an extra weekend to build the children's playhouse, no one will really suffer. But if it takes an additional three months and $2 million to complete a passenger terminal, the consequences are serious. While the playhouse construction manager can afford to be inept, the terminal construction manager does not have this luxury. Nor does any engineer who manages resources on behalf of someone else.

1

Anatomy of a Project

A project is a special kind of activity. It has a beginning, a life, and an end. Thus, it is born, it develops, and it dies. It also has objectives whose accomplishment signals the end, and it has boundaries which it shares with other activities so that its extent is defined.

A project is different from never-ending functions, e.g., the accounting function, the manufacturing function, the sales function, the personnel function, and so forth. Note, however, that these functions may contain projects within them. Projects are also different from activities that have beginnings and ends but no specific goal, such as having dinner, playing the violin, or watching television. And projects differ from programs, whose conclusion is diffuse, and from other activities that have no bounds, such as the sum of all the affairs of a major corporation or a governmental agency. Here too, though, there may be projects within the program or within the total range of affairs.

I. PROJECT LIFE CYCLES

Because a project has a beginning and an end, it has a life cycle. The life cycle starts with a concept phase and concludes with a post-accomplishment phase, as shown in Figure 1.1. Four intermediate phases are also identified in the figure: a project definition (or proposal preparation) phase, a planning and organizing phase, a preliminary studies phase, and a work accomplishment or performance phase. Each succeeding phase is more concrete

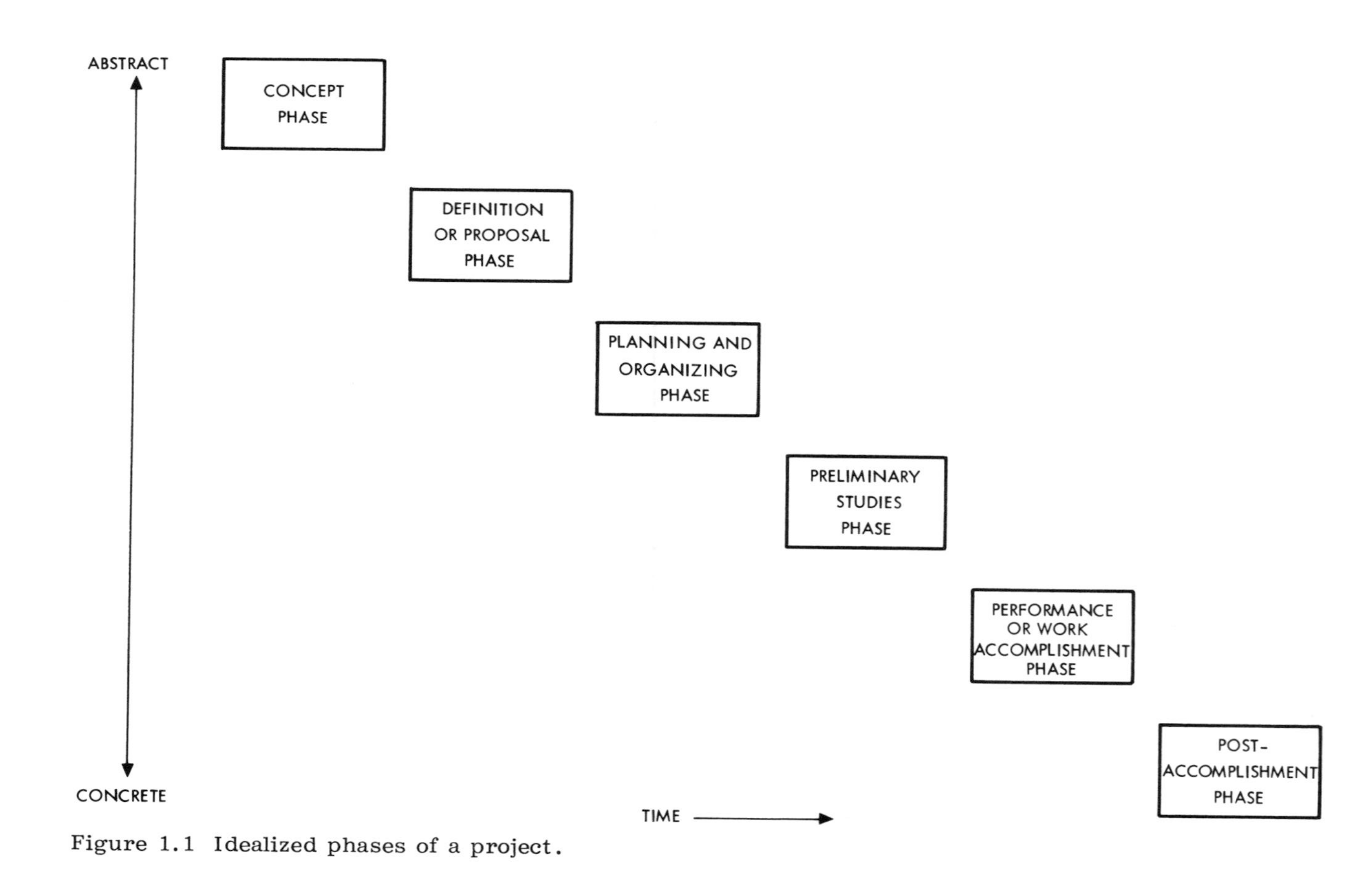

Figure 1.1 Idealized phases of a project.

than the preceding phase, as the project matures from an overall concept to a set of tasks that in their totality accomplish the project.

A. Concept Phase

The concept phase begins with the initial notion or "gleam in the eye" of someone who imagines accomplishing some objective. The objective may be to provide a bridge, to develop a manufacturing capability, to obtain information, to make arrangements to accomplish some task or goal, to build relationships to enable future actions, to win some number of customers, to train some number of employees, and so forth. Sometimes the objective is less specific than, say, to provide a bridge. Perhaps the objective is stated merely as "to provide a means for 1,500 cars to cross the river per hour." This approach is just as valid as specifying a bridge, and it may be more useful because it does not rule out choices that may be better. The only restriction on defining an objective is that one must be able to tell when it has been attained.

It is important to note that the concept phase does *not* include how to accomplish the objectives—such considerations duly come later. Many projects, however, experience grave difficulties because the concept phase is truncated before it is finished and attention is prematurely turned toward the means for accomplishing the objective. A project's objectives need to be fully explored and developed in the concept phase if their means for accomplishment are to be optimized. Otherwise, much time will be spent in needless arguments about the approach because different people will have different ideas about what is to be accomplished.

The concept phase not only includes formulation of project objectives but also identification of relevant constraints. This is not to say that the list of constraints must be exhaustive and that further constraints will not be identified. Indeed, it is a rare project where such foresight exists. Sometimes, however, an objective must necessarily be accomplished within certain limits, such as budget or time, or must be accomplished using certain tools, personnel, or procedures. In these cases, the constraint is so important that it needs to be stated along with the objectives. Otherwise, the project could conceivably be developed in a way that violates a cardinal limitation.

Formulating objectives fully during the concept phase is a major help toward an efficient and relatively peaceful project. It does not guarantee, however, that they will not later be reviewed or reconsidered. As the project matures, the customer* may change its objectives or the project's activities may produce information showing that the objectives are not fully appropriate. In either case, the project should return to the concept phase to confirm or change the objectives. If the objectives are changed, everything else that follows is also subject to being changed!

When the concept is developed to the point that it can be meaningfully discussed and it can be concluded that it has a reasonable chance of solving or resolving the problem at hand, the project is ready to advance to the next stage, project definition.

B. Project Definition Phase

Project definition (and/or proposal preparation) follows the concept phase. This phase has two parts: The first comprises characterizing the project in terms of assumptions about the situation, alternative ways of achieving the objectives, decision criteria and models for choosing among viable alternatives, practical constraints, significant potential obstacles, and resource budgets and schedules needed to implement the viable alternatives.

The second part consists of tentatively selecting the overall approach that will be used to achieve the objectives. Obviously, not everything that eventually needs to be known in order to accomplish the project is known at this early stage. Thus, many choices are made tentatively, with contingency arrangements identified in case the choices are found to be unsatisfactory.

If the amount of uncertainty is so great that the contingency allowances are unacceptably high, two remedies are available. First, the project can be divided into two sequential subprojects. The objective of the first subproject is to obtain information that will reduce the amount of uncertainty. This information is then incorporated in a second subproject directed toward achieving the

---

*Depending upon the situation, "customer" may mean (1) the client's representative, (2) the client's own customer, (3) a group of people in the client's organization, (4) the project manager's boss, (5) another individual in the project manager's own organization, or (6) a group of people in the project manager's organization.

main objective. Second, one can merely pursue the several alternatives in parallel until it becomes clear which to continue and which to abandon. This second approach will generally be more expensive than the first, but it *may* take less elapsed time to reach the ultimate objective.

The project definition produces a plainly written, unambiguous description of the project in sufficient detail to support a proposal or request to the customer to undertake the overall project. The definition should address:

1. how the work will be done
2. how the project will be organized
3. who are the key personnel
4. a tentative schedule
5. a tentative budget

The aim is to convince the customer that the doers know what to do and are qualified to do it. At the same time, the description should not have so much detail that the project is essentially planned or completed even before it is authorized.* Thus, the project definition phase represents a beginning-to-end thinking-through of the project but does not accomplish the project in and of itself. It is like a map of a route rather than the route *per se*, and it is a coarse map at that.

If at all possible, key personnel of the would-be project should be involved in defining the project and preparing the proposal. Sometimes this ideal arrangement is not possible, for legitimate reasons such as unavailability at the moment. Whenever they are not significantly involved at this stage, money and time should be allowed before progressing to the planning phase for them to review the work of this phase and perhaps modify it or convert it to something they can support.

---

*One always has the problem in preparing project proposals of deciding how much of the actual project to do in order to show that it can in fact be done. In general, the answer is that as much should be done as necessary to obtain project authorization, but no more. The exact amount is a matter of judgement and depends upon a variety of factors such as the novelty of the project, the customer's familiarity with the subject and with the project team, and the competitive situation.

C. Planning Phase

Assuming that the customer accepts the project proposal and authorizes the project, the next phase is to plan and organize it. It is in this phase that *detailed* plans are prepared and tasks identified, with appropriate milestones, budgets, and resource requirements established for each task. Some project managers try to do this work during the project definition phase, partly to demonstrate that they know how to manage the project, and then they skip lightly over it once the project is authorized. While the desire to economize is commendable, it is not always effective. Indeed, such efforts are seldom sufficient to actually manage the project, and the overall costs are invariably higher than if proper planning is done. Moreover, the project's scope of work may be revised between the time it is proposed and the time it is authorized, thereby invalidating some parts of a preliminary plan.

This phase also includes building the organization that will execute the project. While some consideration of how the project will be staffed is undoubtedly given during the project definition phase, there is typically no guarantee at this early time that the project will actually be authorized and that the individuals can be committed to the effort. Furthermore, plans are most effective when they are developed by the doers, and doers are best motivated by having a say in planning their work. Therefore, it behooves project managers to develop their organizations and their plans simultaneously, in order to have each enhance the other.

To organize the project team, the project manager must identify the nature, number, and timing of the different skills and traits needed and arrange for them to be available as required. These requirements include not only various sorts of technical expertise, but also skills and traits in such areas as communication, leadership, followership, conceptualization, analysis, detail-following, initiative, resourcefulness, wisdom, enthusiasm, tolerance for ambiguity, need for specific information, and so forth. No single prescription can be given for all projects in terms of what is required. Suffice it to say that there should be an appropriate range and mix of the characteristics needed for the project to succeed.

Much more is said about project planning in Chapter 3.

D. Preliminary Studies Phase

Once the project has been planned and organized, it's time to begin doing it. Haste makes waste, however, and most project managers will help themselves considerably by deliberately beginning the work with "preliminary studies." The preliminary studies phase consists of literature searches, field reconnaisances, experiments, interviews, and other forms of data or information gathering which first of all validate or rectify any assumptions made in the plan and secondly identify and characterize critical aspects of the project so that it can go forward smoothly.

Most features revealed by preliminary studies will show up later if the studies are not performed. Discovering them late, however, is likely to be disastrous or embarrassing. It will be disastrous if it turns out that key assumptions are invalid and that the remaining time or resources are inadequate to achieve the objectives. And it may be embarrassing if it turns out that time or resources have been needlessly squandered. Sometimes a discovery is both disastrous and embarrassing.

Insight, candor, and a realistic outlook on the part of the project team are necessary to define appropriate preliminary studies. Insight is needed to identify the project's areas of vulnerability, to identify where preliminary studies are most useful. Candor is needed to be able to voice these concerns. And a realistic outlook is needed to keep from dismissing them as insignificant. If any one of these ingredients is missing, the project stands a good chance of disaster or embarrassment. Project managers do not need such burdens, so it behooves them to pay proper attention to preliminary studies.

If preliminary studies confirm the team's fears about being successful, they should revisit the concept, definition, and planning phases, quite likely in consultation with their customer. If the studies, however, indicate that success is within reach, it is time to proceed to the performance phase.

E. Performance Phase

The performance or work accomplishment phase is the part of the project that most people think of when they think of a project. Basically it consists of doing the work and reporting the results. Doing the work includes directing and coordinating other people and controlling their accomplishments so that their collective efforts achieve the project's objectives. These topics warrant entire chapters and are discussed in Chapters 5 and 4, respectively.

F. Post-Accomplishment Phase

It is commonly believed that completing the last task in the performance or work accomplishment phase concludes the project, but there is yet a final phase called a post-accomplishment phase. This phase consists of:

1. confirming that the customer is satisfied with the work and performing any small adjustments and answering any questions necessary to achieve satisfaction
2. putting project files in good order so that they will be useful for future reference
3. restoring equipment and facilities to appropriate status for others to use or decommissioning them
4. assuring that project accounts are brought up-to-date, appropriately audited, and closed out
5. assisting project staff in being reassigned
6. paying any outstanding charges
7. collecting any fees or payments due

G. Phase-to-Phase Relationships

While Figure 1.1 and the preceding discussion tend to make the interphase boundaries distinct, the successive phases often overlap in time. As the project is defined, for example, some second thoughts about the concept may arise and it will be revised. Thus, the concept phase is extended so that it overlaps the project definition phase. Such overlaps are shown in Figure 1.2.

In addition to phases overlapping, it is likely that some iteration among, say, the preliminary studies phase, the concept phase, and the project definition and planning phases will occur. Certainly there are cases where the results of the preliminary studies have caused the entire project to be revamped.

II. PROJECT BOUNDARIES, INPUTS, OUTPUTS, AND INTERFACES

A. Project Boundaries

Project managers should recognize the boundaries of their domains. Elements that lie within a project's boundaries are clearly the

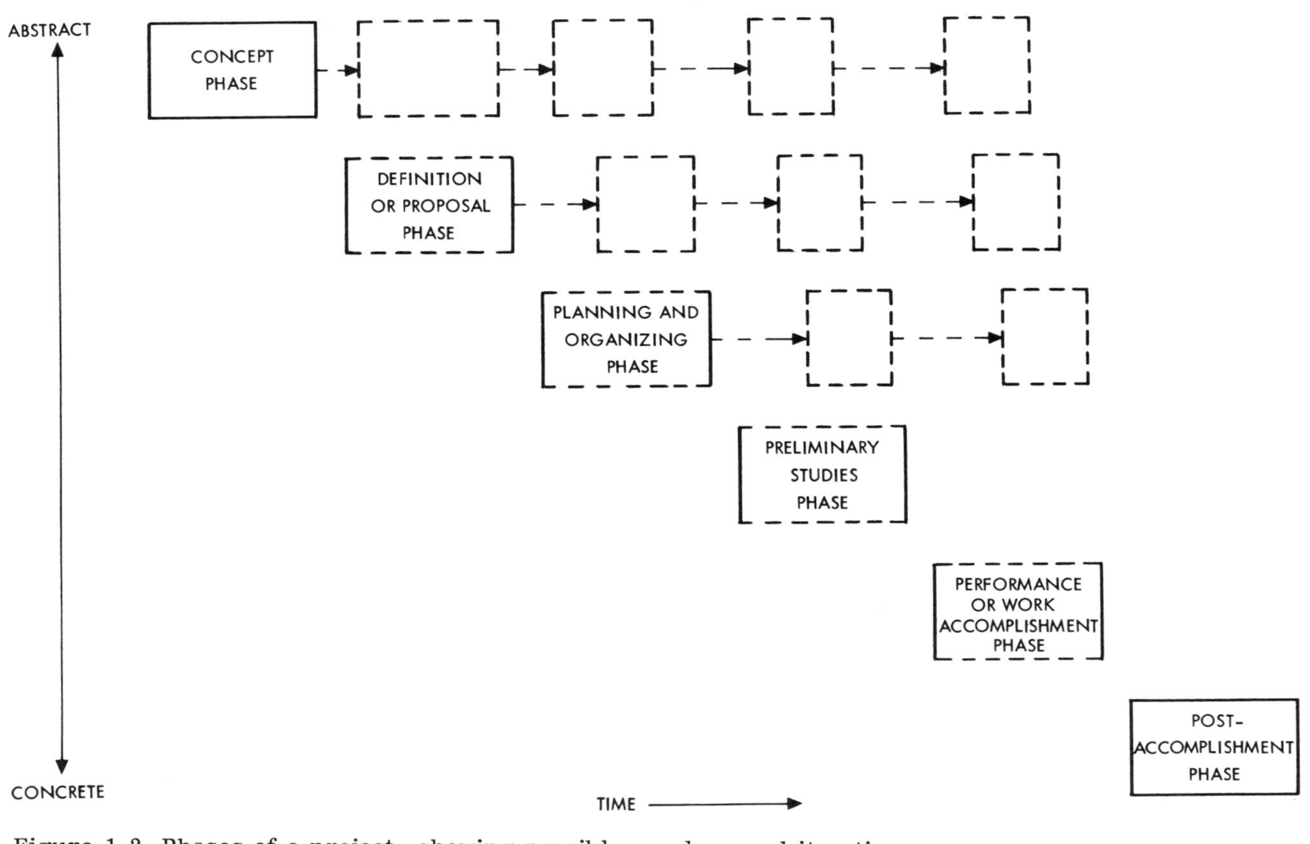

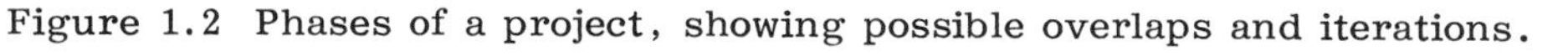

Figure 1.2 Phases of a project, showing possible overlaps and iterations.

responsibility of the project manager, who has some measure of authority over them. Elements that lie outside the project's boundaries are subject only to the manager's influence; by definition they are not the manager's responsibility.

While information and resources must generally flow back and forth across project boundaries in order for the project to succeed. project managers may have very limited direct control over these vital exchanges. Accordingly, they must use their insight and diplomatic and persuasive powers to ensure that their projects are not strangled by inappropriate flows. That is, project managers must cause flows to happen without being able to command that they happen. Many times this task requires project managers to negotiate with their counterparts on the other side of these boundaries. The art of negotiation is so important in project management that we devote an entire chapter to it, Chapter 7.

B. Project Inputs and Outputs

The information, materials, and resources that flow into a project are called inputs, and the information, materials, and resources that flow from a project are called outputs. These items are described below.

1. Project Inputs

a. Scopes of Work Scopes of work are statements of project objectives and major constraints, including schedule and budget. Sometimes they are quite detailed; other times they are so poorly stated that the project cannot proceed until they are refined. But in any event, the scope of work is a major input to the project.

A scope of work may originate with the customer, or it may originate with the organization that will perform the project. The latter may seem strange, but often the would-be customer asks a would-be project organization to propose a project. Although the project organization may have proposed the scope of work to the customer originally, it is only a proposal at this early stage. When it comes back as a contract, it is really the customer's. No project manager should think that the scope of work can be changed without obtaining the customer's prior approval, even if the project organization submitted it as a proposal in the first place.

The project manager should also recognize that work statements described in proposals are often modified by the customer unilaterally or via negotiation with the would-be project organization before they are issued for performance. It behooves project managers to read, nay study, their scopes of work before beginning their projects. And they should keep them handy for future reference (or memorize them). Otherwise, they are prone to working on their projects under false conceptions.

b. Contract Terms Contract terms are also project inputs. They may be either innocuous or consequential, depending upon how well they describe how the project must proceed and the mutual responsibilities of the customer and project organization. Often they are written in a standard format by the customer and applied indiscriminately. Sometimes standard terms are unfair or unsatisfactory to the project organization because of the project's nature. In these cases, they should be revised or removed through negotiation *before* the contract is signed. Also, certain projects need contract terms that are non-standard and that will appear in the contract only if the project manager sees to it that they do. Project managers should study the terms of their contracts for the work they will do in order to ensure (1) that contract terms will not so hinder them that they cannot perform their projects, and (2) that contract terms will induce the other parties to behave in ways that will aid them in performing their work.

Among the contract elements that are of concern to a project manager are:

1. a change clause, i.e., a provision that deals with how changes to the scope of work are to be handled
2. a clause on allowable costs, i.e., the provision that tells the manager what costs the customer will pay for
3. a clause regarding giving of notice in the event of schedule or budget overruns
4. a clause on subcontracts and the selection of subcontractors
5. a clause on excusable delays
6. a clause concerning the customer's provision of information and materials and any sanctions for tardy delivery
7. clauses dealing with substantial compliance and deficiencies
8. clauses dealing with termination for convenience and for default

The above discussion on contracts implies that project managers are performing work for an external organization, but this need not be the case. Even when projects are done for one's own organization, project managers need to work under conditions that help rather than hinder their jobs. Thus, even here project managers need to negotiate the terms and conditions of their assignments before agreeing to take them.

c. Organization Policies Organization policies are important inputs in most projects because they guide the way work is performed. For example, they may mandate certain types of reviews and approvals before work proceeds from one stage to the next. They may specify how subcontracts are handled; how support services, e.g., typing and accounting, are provided; how personnel are assigned; and how all-important client contacts or other relationships are maintained. Sometimes they are written down; often they are not. They are nevertheless important influences on project managers' behavior, and they need to know how to succeed in their presence.

d. Project Personnel Among the most important inputs for project managers are the personnel who are assigned to their projects. They bring technical knowledge and skills, interests, aptitudes, and temperaments. Most projects need a mix of these characteristics, as described in Chapter 5. The project manager should try first to identify the kinds of talents needed and then to secure them. It is usually not sufficient merely to tell the personnel department that the project needs two X-ologists, one Y-ologist, and four Z-ologists.

e. Material Resources Material resources comprise the facilities and equipment that can or must be applied to the project. Often they limit how the project can be achieved. Or, if they have unusual capabilities or versatility, they may enable particularly effective or efficient approaches which are not always used. In either case, project managers should understand how the material resources available to them are likely to affect or facilitate their projects. They should be identified as to:

1. what the customer will furnish
2. what can be obtained from within the project manager's own organization
3. what shall have to be obtained elsewhere, including likely sources

f. Information Finally, we list information as a project input. Information may be technical, economic, political, sociological, environmental, and so forth. It can come from other members of the organization, from customers, from sub-contractors and vendors, from third parties such as governmental agencies, and of course from both open and restricted literature.

The quality and quantity of information which the project organization obtains, either on its own initiative or on the initiative of others, will have major impact on the amount and type of work that can be and must be done. Project managers and their teams should cultivate their information sources and use them advantageously. Little else that they can do will be as cost-effective if the information exists.

2. *Project Outputs*

A project can be visualized as a processing machine. The inputs from Section 1 are processed by the project activities to produce the project outputs:

1. deliverables
2. internal information
3. experienced personnel
4. working relationships

a. Deliverables Deliverables are the visible products of the project, the items that the project is supposed to produce. They may include tangible items such as hardware or a structure, or information such as instructions, an analysis, a report, drawings and specifications, a design, contract documents, or a plan.

b. Internal Information Internal information consists of the increased knowledge base that the project staff generate as a result of doing the project. This information can be invaluable to the organization in a future project, whether for the existing customer or another customer. It may consist of information about the customer, about subcontractors, about production conditions, and about the environment as well as about the project itself. The information may be in the form of memoranda, report drafts, project instructions, project standards, check prints, and so forth.

Since the project is not required to deliver such internal information, it is often neglected. Astute project managers, however, will capture and retain such information whenever they can in order to build their knowledge base for future assignments.

c. Experienced Personnel A major output of every project is experienced personnel. Whether the experience is in fact beneficial or harmful depends of course on how well the project is conducted. If it is conducted satisfactorily or even outstandingly, project staff will grow and develop and generally enjoy their experience. They will be both more skillful and willing to work with the project manager in the future, which is a plus for the project manager. On the other hand, if the project is conducted poorly, project staff will probably not enjoy their experience even if they do grow and develop. The odds in such circumstances are that the staff will prefer not to work with the project manager if they can avoid it, which limits the manager's options, if not his or her effectiveness, in the future.

d. Working Relationships Another project output is the set of working relationships that the project manager develops with other internal departments and with external organizations. Like the quality of the project staff's experience, these relationships may be either positive or negative and they will bode either well or ill for the future. Project managers should take care to develop the kinds of relationships on current projects that will serve them well on subsequent projects.

C. Project Interfaces

A typical project has literally dozens of interfaces across which information and deliverables flow. These interfaces are with various departments and offices in the project's own organization, with various functional elements in the client's organization, with different elements in subcontractor and vendor organizations, and with third parties such as regulatory agencies, auditors, and so forth.

A major duty of every project manager is to ensure that all project interfaces function so that necessary information and materials are properly transmitted. The first step in managing these many interfaces is simply to list them and to assign each one to a project staff member who will keep it functioning. To help in this regard, Table 1.1 lists representative interfaces for reference.

The next step is for all project staff members to establish face-to-face or telephone contacts with their interface counterparts and to make their existence known. They should exchange (or at

TABLE 1.1
Potential Project Interfaces

Within the organization

Accounting office

- accounts payable
- accounts receivable
- project accounting records

Administrative or business manager

Computer facility

- application programmers
- supervisor

Contract administration

Drafting department

- drafters
- supervisor

Engineering disciplines

- engineers
- supervisors

Equipment facility

Field offices

- supervisor
- technicians and other staff

Insurance office

Laboratories

- supervisors
- technicians

Legal Department

Library

Main Office

- central engineering
- "inspector general" for projects

Other offices

- office head
- project managers

Other project managers

Personnel office

Printing department

Public relations department

Technical project office

- supervisor
- technical project officer

Technical specialists

Transportation office

With the customer's organization

Accommodations office

Accounts payable

Contract administration

Drafting department

Equipment supply office

Field staff

- supervisor
- technicians

Industrial relations department

Legal department

Public relations department

(continued)

TABLE 1.1 (Continued)

Purchasing department	Secretaries
Receptionist	Security department
Safety department	Subcontract administration
Scientific disciplines	Transportation office
discipline managers principal investigators	Travel office
With subcontractor and vendor organizations	
Accounts receivable	Project manager
Contracts administration	Public relations department
Crew chiefs	Purchasing department
Equipment supply office	Sales department
Insurance office	Storekeepers
Legal department	Technical specialists
Operators and technicians	Transportation department
Pricing department	Workshop
With third parties	
Auditors	Learned societies
Banks	Libraries
Competitors	Media
Data bases	Planning and zoning agencies
Equipment rental firms	Politicians
Express and freight forwarding companies	Pressure groups
Family members	Surveyors
Governmental agencies, including permit issuers	Utilities
Land owners	Tax assessors
	Travel agents

least begin to exchange) information with each of them on their respective needs, wants, and expectations. Each staff member should report back to the project manager the agreements that they and their counterparts have made about the nature, form, and frequency of their future exchanges. The work implied by these future exchanges should then be factored into their project plans, which are discussed in Chapter 3.

As information and deliverables begin to flow across the interfaces, the interface-keepers should verify that they are serving their intended purposes. This responsibility may require some deliberate follow-up activity, for recipients may not report whether the items satisfy their needs and expectations and whether additional work or information would be helpful.

2

Roles and Responsibilities of the Project Manager

Project management involves a set of duties that must be performed and are no one else's prime responsibility. We have organized these duties into six roles and responsibilities. Depending upon the project organization, the first five may be done by the project manager directly or they may be done by others under the project manager's supervision. The sixth one, "inherent duties," can be done only by the *de facto* project manager.

1. ASSURE CUSTOMER SATISFACTION

The single most important responsibility of the project manager is to assure customer satisfaction. If the project is successful in every respect in terms of meeting its stated objectives, schedule, and budget but the customer is somehow not satisfied, then the job was not done well enough. Such dissatisfaction could arise because the customer had a preconceived, but unspoken notion of what the outcome should be, because the customer obtained new information that caused a revision in priorities, or because the customer had a change of mind.

The project manager can learn about and keep abreast of the customer's needs and expectations by practicing some simple procedures:

1. Confirm key issues during the course of the project so that the work done is adjusted to meet current needs and expectations.
2. Develop a friendly intelligence system within the customer's organization to obtain early warning signals of changes in emphasis, priority, etc. and prepare to respond appropriately to them.
3. Reread the contract several times during the project and attend to all its requirements, including administrative aspects, so that no needs or wants are left unfulfilled.
4. Keep the customer informed and up-to-date so that he or she is prepared for the eventual results and has an opportunity to influence the way that they are developed and presented.

If there is a written contract, it may seem strange that it is necessary to be in touch continually with the customer. However, a project situation is seldom straightforward and the contract may not reveal all that needs to be taken into account. For example:

1. The customer may not have stated the requirements clearly or completely. Contact is necessary to determine the true objectives and constraints.
2. The contract or statement of work may not reflect the needs of all the elements of the customer. Contact is necessary first to identify all the interested elements and then to identify their needs.
3. The customer may have constraints or a hidden agenda which cannot be put in the contract or scope of work but whose existence controls the range of useful project results. Contact is necessary to ferret out these unwritten limitations.

The entire project team can develop information about their customer as they do their day-to-day work on the project, and the project manager should guide them in fulfilling this role. Many inexperienced project personnel need to be instructed regarding these duties.

Finally, if it should become clear that the customer cannot possibly be satisfied, then the project manager should:

1. be sure that the contract is satisfied to the letter
2. be sure that the customer pays all invoices due
3. terminate the relationship

II. DIRECT AND CONTROL ALL DAY-TO-DAY ACTIVITIES NECESSARY TO ACCOMPLISH THE PROJECT

A project does not run itself. Someone must direct and control it, and that someone is the project manager. The project manager is responsible for directing and controlling all the day-to-day activities that are necessary to accomplish the project.

This responsibility does not mean that project managers cannot delegate the direction or control of specific activities to others. On the contrary, they may indeed delegate, but the *responsibility* for the direction and control is still theirs. If the delegates do not perform adequately, the project managers are responsible for the inadequate outcomes and for repairing any damage done.

Good project managers arrange matters so that their teams avoid troubles and have appropriate back-up plans and resources for risky situations. By their foresight, their projects run more-or-less smoothly. The more successful they are in this regard, the less likely anyone will be aware of their effectiveness. As a result, they may not be fully appreciated. Only sophisticated observers will recognize the high quality of their efforts.

A project manager may find from time to time that a boss, a salesperson, a functional manager who supplies some staff, or even a customer is trying to direct part of the project. When this happens, the project manager must firmly rebuff the interference. A project manager may well listen to advice from these people but should not let them direct project staff members. Otherwise, there is a grave risk that the project will be seriously damaged overall. This occurs because the would-be intervenor invariably lacks understanding of the entire project and its interdependencies. While tinkering with one part, the intervenor upsets the project's overall balance even if well-intentioned.

Directing and controlling day-to-day activities is both time-consuming and sporadic. Its sporadic nature makes it difficult to schedule completely. If project managers are not careful, they may leave too little opportunity in their own work plans to accommodate the many conversations and informal meetings that are necessary to direct and control their projects. As a rule of thumb, they should commit to formal meetings and appointments at most half of their time. The other half will be filled with directing and controlling day-to-day activities.

III. TAKE INITIATIVES AS REQUIRED IN ORDER TO ACCOMPLISH THE PROJECT

The project manager is the chief initiator on the project. From time to time he or she will face and have to resolve problems that no one could foresee. There will always be revolting developments and they must be expected. Action is required when they happen.

In all likelihood, project managers will not be able to resolve all revolting developments alone. They may need to consult an expert but not know to whom to turn. When this happens, they can call the one person whom they think is most likely to know more about the problem than they do. The chances are that this first contact will not be able to provide all the help needed but can probably give one or more leads to people who are more expert. They can continue this chain a few more times. The chances are very good that by the fifth phone call they will be connected to someone who will be a suitable expert for the problem. (This process is known as the "five phone-call phenomenon.")

If the bottom drops out and the project manager gets a sinking feeling, he or she should not just stand around and act like a bump on a log. That would only cause others to do so, too. The project manager must take the initiative and attempt to solve the problem as quickly and as orderly as possible.

IV. NEGOTIATE COMMITMENTS

The project manager is responsible for negotiating the commitments that the project team will perform for their customer. As explained in Chapter 7, this does not necessarily mean that the project manager is an active negotiator for all the commitments, but he or she is responsible for the team's negotiations. In the end, the project manager is responsible for fulfilling whatever is negotiated, and therefore has a vital interest in the outcome of the negotiations and must be in agreement with it. If not, then the project manager is not prepared to manage the project.

Sometimes a project manager is assigned to a project after some commitments have already been made to the customer. The project manager's first duty in these circumstances is to evaluate the situation and determine if the project team can meet the commitments. If they can, well and good. If they cannot, then the project manager must tell management and/or the customer what problems exist and negotiate relief from unreasonable objectives and constraints. If the commitments cannot be changed, then the project manager should document the envisioned difficulties to management so that they will know what to expect.

In most projects, negotiations continue during the work. Most customers recognize that nothing is certain and that changes may be necessary once the facts of the situation emerge. Depending upon the degree of specificity of the project/customer contract or arrangement, formal changes in objectives, schedule, budget, staffing, strategy, and methods may have to be negotiated.

Negotiating is discussed further in Chapter 7.

V. ENSURE COLLECTION OF THE FEE

The project manager is responsible for assuring that any fees due are collected from the customer. In some organizations it is efficient and effective for this task to be handled by someone other than the project manager. However, the project manager should always know how well the customer is meeting the contractual commitments of payment. If payment should lag seriously, the project manager may stop work on the project, both to avoid incurring further expenses without reimbursement and to induce the customer to pay for the work already done.

VI. INHERENT DUTIES

Below are listed nine duties that we believe are inherent duties of the project manager. Putting it another way, the *de facto* project manager is the one person who performs *all* these duties on a project.

If these duties are divided among two or more people, then the project manager's role and responsibility are also divided. This means that the project staff and all others who deal with the project will be confused about who the project manager really is. Also, the "alternative" project manager will be consulted and notified when the nominal manager should be. The staff can thus receive instructions contrary to those that the project manager would give and he or she will lack important information needed to manage the project.

A. Interpret the Statement of Work to Supporting Elements

Interpreting the statement of work is the first opportunity for the project manager to show leadership and is a critical step. When done well and timely, it is a good motivating tool for the entire project team. It is the key to everyone on the team having a common understanding of the objectives, constraints, and major interactions on the project.

B. Prepare and be Responsible for an Implementation Plan

The project implementation plan must be the project manager's plan. Even if others help in putting it together, there should be no mistake that it is *the project manager's* plan. The project manager and everyone else should consider a violation of the plan as a violation of the project manager.

If someone becomes the manager of a project in mid-course, then the new manager should develop his or her own plan and publish it promptly. Even if the plan left by the previous manager is acceptable and accepted, the new manager must affirm that it is his or her plan and commit to it.

C. Define, Negotiate, and Secure Resource Commitments

The project manager should define, negotiate, and secure commitments for the personnel, equipment, facilities, and services needed for the project. These commitments should be as specific as possible in order to verify first that they are appropriate, and second that they are being kept. Through this activity the project manager establishes his or her role in the eyes of all those who provide resources and support to the project.

D. Manage and Coordinate Interfaces Created by Subdividing the Project

As explained in Chapter 3, projects are typically broken into subprojects that can be assigned to individuals or groups for accomplishment. Whenever the main project is broken down this way, the project manager must manage and coordinate the high-level interfaces that are formed by subdividing the work. (Intermediate and lower level interfaces are managed and coordinated by the task leaders who create such subdivisions.)

Interface management and coordination consist of making sure that each leader of a subproject provides whatever another subproject leader needs and does so in a timely way. This may require intervening in a subproject leader's plan, since meeting interface requirements may not be the leader's own highest priority task. Likewise, the project manager may need to persuade another subproject leader to moderate his or her requirements because meeting them may place undue strain on the subproject providing the needed items or information.

Sometimes subproject interfaces can effectively be managed and coordinated by holding a series of separate meetings with the individual subproject leaders. At other times, however, best results are obtained by having all the subproject leaders discuss their needs and constraints together, at a project meeting. A project meeting enables all the participants to contribute toward resolving their common difficulties without their views being filtered through the project manager. Moreover, it generally facilitates imaginative solutions and a full discussion of side effects, which are less likely to occur in a series of individual discussions.

It is best all the way around if the work is broken so that interface communications and activities are minimized and as straightforward as possible. It is usually easier for a single individual or unit to bring order out of chaos than for two individuals or two groups. Accordingly, the work should be broken along organizational lines wherever possible. This will minimize the amount of interface management and coordination required of the project manager and help make available the time needed to do other tasks.

E. Monitor and Report Progress and Problems

The project manager is responsible for reporting progress and problems on the project to the customer, the boss, and all others who need to know. This responsibility should not be delegated lest the outsiders start to go directly to team members with their concerns.

Team members should be instructed to refer inquiries and requests from outsiders to the project manager if they involve issues not directly within their own personal control. Also, when team members do respond to outside inquiries or requests they should promptly notify their project manager of their actions.

In order to report progress and problems in a timely way, project managers must monitor their projects and not merely wait until others bring them news. They must be aware of what is going well and what is going awry and should seek these kinds of information from their project staff. They may have to persist in uncovering unhappy news, for many people are reluctant to give it. If, however, they explain that they prefer early warnings to late surprises, help their staffs surmount their difficulties, and do not shoot their messengers, then they stand a good chance of getting the information that they need first to keep their projects on course and second to keep outsiders properly apprised of their situations.

Details of the process of monitoring are discussed in Chapter 4 on control techniques.

F. Alert Management to Difficulties "Beyond One's Control"

Occasionally a project manager finds that the only way to relieve a bad situation is to get help from outside the project. When this happens, the project manager has a difficulty "beyond one's control." Suffering the difficulty in silence will not be rewarded. The project manager's management or customer must be alerted so that other resources can be brought to bear, constraints can be relaxed, or project objectives can be adjusted.

It is not sufficient just to mention the difficulty in passing. The discussion must be an overt act and should be confirmed in writing. If the management or customer says that they will take specific action to resolve the difficulty, that too should be in writing, together with the time when they say they will do it.

G. Maintain Standards and Conform to Established Policies and Practices

The project manager must set and maintain the standards that will govern the project staff members. Where pertinent policies and practices have already been established, the project manager is responsible for seeing that the team conforms to them.

Whether or not project managers take overt action, most project staff will take their cues from them regarding acceptable standards of behavior and performance. The occasional maverick who resists following these norms should be faced directly on issues that are important in order to build a group or team approach to the project. The performance of the regulars can be undermined by allowing a maverick's performance or behavior to go unchecked on an important issue.

At the same time, project managers should not impose needless constraints or standards. Nor should they act in ways that appear arbitrary or capricious. If they are to win and maintain the respect of their team members, their actions should be seen as helpful and reasonable, not handicapping or highhanded.

H. Organize and Present Reports and Reviews

The project manager is responsible for organizing and presenting reports and reviews to the customer and to management. He or she is the "focal point" for the team and should be seen as such.

This duty does not imply that project managers must either prepare or present 100% of their project reports or reviews. But they are responsible for them and should orchestrate them, send written reports under their cover letters, and be the first and last speaker and "the glue" that holds a review together.

The project manager should not be intimidated by members of the audience who ask questions that only expert staff members can answer. When such questions arise, as they invariably do, the manager should turn to appropriate staff members if they are present or promise to get a reply back to the questioner if they are absent. The project manager is not expected to know every last detail, but is expected to have the resources available to handle pertinent issues. (Impertinent questions should be politely rebuffed on the spot.)

I. Develop Personnel As Needed to Accomplish the Project

It is a rare project indeed where all the talents needed to do the work are present in the staff available. The project manager must thus train them and compensate for their shortcomings. A project manager may, for example, have to teach staff members how to plan their work, interface with others, report their results, and generally how to function as a project team member. The manager may also have to teach them how to practice their specialties under adverse circumstances by showing them how to provide for risk.

Whereas project managers can delegate parts of the tasks described in Sections I through V, they dare not delegate their inherent duties. To do so is to confer upon others part or all of the role of project manager and undercut their own ability to manage. In this event, they will be lucky if their projects turn out successfully.

If one is assigned as a project manager and cannot perform all nine inherent duties, he or she should rebel! If one is asked to take over a project in trouble and cannot perform all nine inherent duties, he or she should rebel! Perhaps this explains why the project floundered in the first place.

3

Planning Techniques

It was mentioned in Chapter 2 that an inherent duty of a project manager is to prepare and be responsible for an implementation plan. This duty lies at the heart of project management and warrants a chapter by itself. The duty includes defining tasks, making estimates, and preparing schedules and budgets. It also involves assigning overall segments of the plan to individuals, with a minimum of interface problems.

I. WHY PLAN?

Before embarking on the discussion of planning techniques, it is appropriate to address the question, "why plan?" Isn't the plan outlined in the project definition or proposal adequate? In general, the answer is that the project definition or proposal plan is incomplete and too superficial to serve as a project *management* plan. It is typically prepared to sell the project and does not address many elements needed to manage the project. And it may be a "success oriented" plan, which is hardly a protection against mishaps or a means of contending with them. Thus, the project manager must develop a project plan in sufficient detail, which is discussed later.

There are still additional reasons to plan:

1. The plan is a vehicle for discussing each person's role and responsibilities, thereby helping direct and control the work of the project.
2. The plan shows how the parts fit together, which is essential for coordinating related activities.

3. The plan is a point of reference for any changes of scope, thereby helping project managers deal with their customers.
4. The plan helps everyone know when the objectives have been reached and therefore when to stop.

II. DEFINING TASKS AND WORK BREAKDOWN STRUCTURES

The first step in developing an implementation plan is to prepare lists of overall project objectives and subsidiary requirements, e.g., reporting and billing requirements. These can be determined by rereading the scope of work and contract as finally negotiated and by referring to organizational policies and procedures. Together these items form the starting point for defining the tasks that will constitute the implementation plan.

The items identified as objectives and subsidiary requirements should then be examined to determine if any of them are hierarchically related* or if there is any overlap among them. If so, they should be reformulated to avoid the overlap and to make the hierarchical relations clear.

The main activities that are needed to accomplish the reformulated objectives and requirements can then be displayed, together with their subparts, in a chart such as the one shown in Figure 3.1. This chart is for a feasibility study and preliminary design for a water supply line and supporting facilities. The chart resembles an organization chart and is the beginning of a work breakdown structure, which will be described shortly.** The main activities are shown at Level I and the subparts are shown at Level II. It is

---

*One element is said to be hierarchically related to another if one of them is a part or subpart of the other.

**Some project managers merely make lists of tasks or work items to be done instead of developing a work breakdown structure. However, this approach is relatively incomplete: It may not address the assignment of responsibility, it may not specify interactions or interfaces required, and it may omit elements of control that will have to be provided later. The work breakdown structure is the most straightforward way so far developed for planning all these aspects. Further, it is no more complicated, and is indeed a lot less complicated, than trying to do all the pieces separately. Once project managers have mastered the work breakdown structure approach, they will know when it is appropriate to take shortcuts.

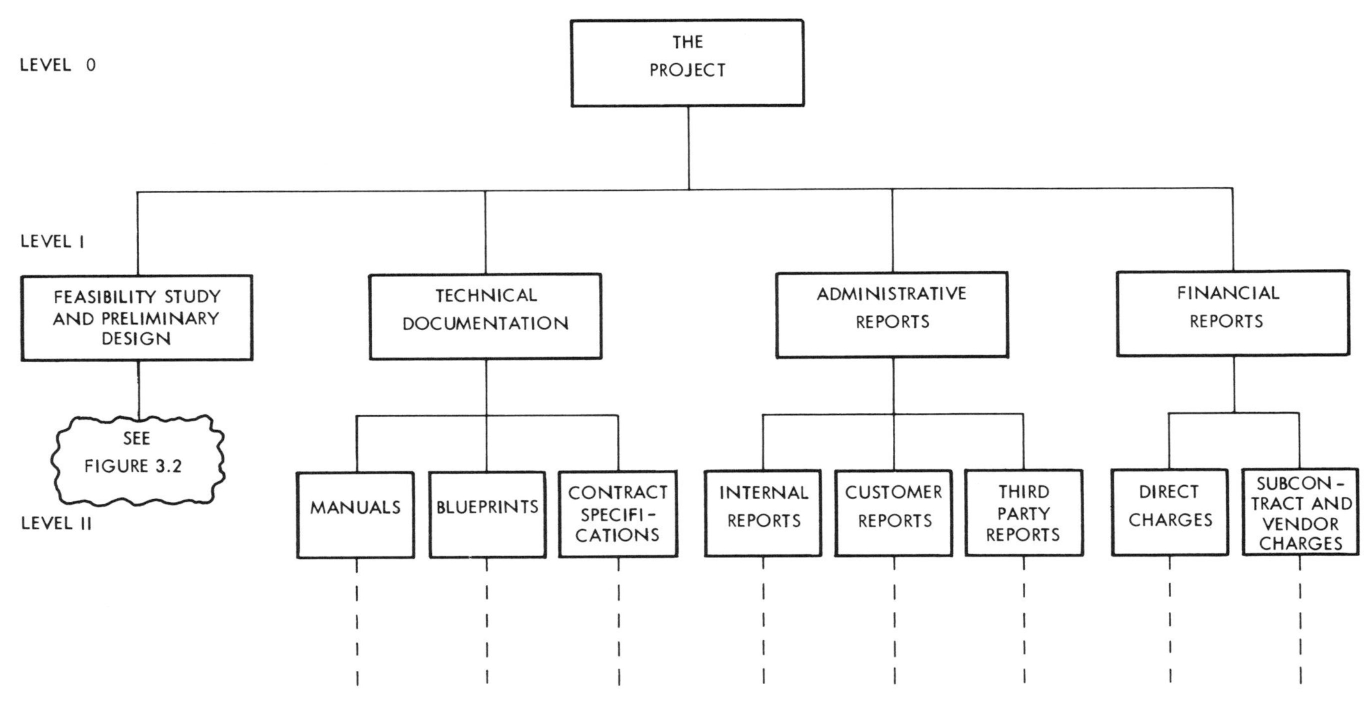

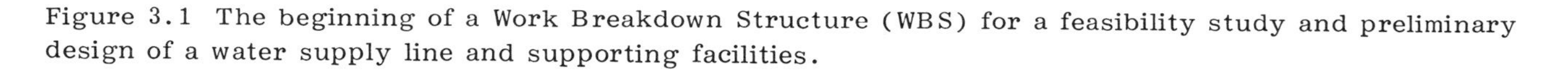
Figure 3.1 The beginning of a Work Breakdown Structure (WBS) for a feasibility study and preliminary design of a water supply line and supporting facilities.

important in preparing this chart that elements that are hierarchically related be shown that way and not shown as parallel parts. Likewise, elements that are truly in parallel should be shown in parallel and not hierarchically.

After the main activities and their subparts have been determined and put on the chart, each of the latter is analyzed into its component parts. Figure 3.2 shows these components for a portion of the feasibility study and preliminary design represented in Figure 3.1. To keep the drawing to one page, the breakdown of only one element is shown at each level. A full work breakdown structure contains the breakdowns for every element at each level and requires the equivalent of many pages. Again, care should be taken to represent accurately the hierarchical and parallel relations that exist among the elements.

The project manager may need assistance from project staff members in analyzing the major work items into appropriate subparts. This is both natural and useful, for it facilitates conferring with the staff about their respective assignments. Indeed, one of the virtues of this approach to defining tasks is that it leads to specific assignments to specific groups and eventually to specific individuals.

While the project's initial objectives and subsidiary requirements and the main activities needed to accomplish them may be quite broad and extend beyond the scope of any existing functional group or individual, eventually the breakdown results in work packages that do correspond to existing groups. Work at these levels can be delegated to the groups, who then can manage all the efforts that comprise the delegated assignment. In the end, the work is broken down into elements that are small enough to be assigned to individuals.

In all but the simplest projects there is usually some discretion about how the larger work elements are to be broken down into their subparts. Generally, confusion is minimized and efficiency and effectiveness are enhanced by observing the following guidelines:

1. Work units should be partitioned so that the subparts that eventually result can be assigned intact to existing organizational units. Such partitions should be made so that the amount of coordination needed across organizational boundaries is minimal and straightforward.
2. Work items should eventually be broken down into elements that are so detailed that the person assigned to each one

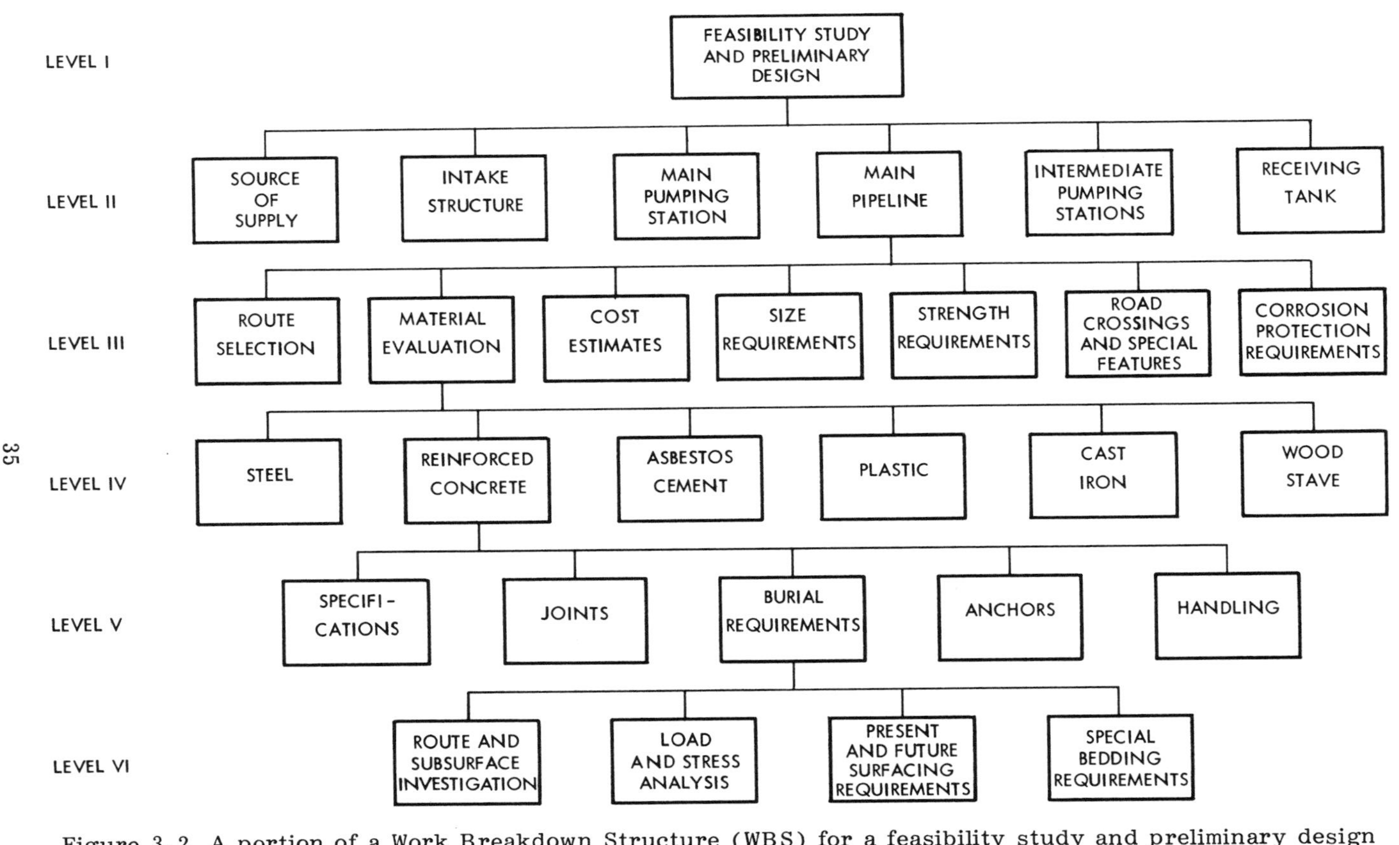

Figure 3.2 A portion of a Work Breakdown Structure (WBS) for a feasibility study and preliminary design of a water supply line and supporting facilities.

can answer the following questions authoritatively and unambiguously:

a. Has the element been completed?

b. If the answer to (a) is "yes," was the effort successful?

This approach means first of all that the various blocks in the work breakdown structure are each assigned to individuals who are competent to judge the completeness and success of the tasks assigned to them. And this approach means that tasks are assigned in such a way that these individuals need to check only with those who are responsible for hierarchically-related subparts in order to answer the questions.

Associated with each element in a work breakdown structure, regardless of its level, are its inputs and outputs. The outputs of one level are the major inputs of the next higher level. This relationship follows from the way that work is subdivided and responsibility is assigned. When a work element is subdivided into smaller pieces, the task of coordinating the resulting interfaces and assembling the pieces to form the larger element remains with the person who subdivided the element in the first place.

Some elements may require inputs from other elements that are not hierarchically related, i.e., from elements that are in a parallel branch of the work breakdown structure. For example, in Figure 3.2, the designer of the main pumping station needs information about pressure losses that depend upon the materials selected for the main pipeline. The person responsible for the element that requires the input from a parallel branch must arrange for it to be provided in the same way that he or she must coordinate inputs from elements that are hierarchically related.

In all likelihood, there will be some high level work items that transcend existing organizational units. These items are the project manager's responsibility. If they are few or small, and within his or her expertise, the project manager may perform them personally. Alternatively, the project manager may assign them to other individuals or to *ad hoc* groups created for the purpose. If any of the members of these *ad hoc* groups are also involved in the project in other ways, it is important that they keep their various roles separated. This is especially important if any of their different roles are hierarchically related. Otherwise, they may be tempted to not manage the lower level tasks carefully because they are in effect reporting to themselves.

From the preceding discussion, it follows that each element of a work breakdown structure not only represents an activity but also a set of inputs and outputs and a responsible individual or group. Developing these details is time-consuming, but it should not be treated lightly. The success of the project depends greatly on making these factors explicit.

Further comments will be made about how detailed the lowest level tasks should be, but they will have to await discussion of control techniques (Chapter 4).

III. PRECEDENCE CHARTS

A work breakdown structure shows graphically how higher level tasks are decomposed into more elemental tasks, but it does not show any chronological relationships among the detailed tasks. For this latter purpose, a precedence chart is used.

A precedence chart is a flow diagram that shows the sequence of both individual tasks and coordinating tasks for the entire project. In particular, it shows sequential tasks in a head-to-tail relation. One method of representation is to draw a series of boxes or circles, one for each task, and to join them by arrows to show the sequence in which the tasks must be performed. An alternative method of representation, and the one used here, is to draw a series of arrows, one arrow for each task, which are themselves joined head-to-tail to show the sequence in which the tasks must be performed. Figures 3.3(a) to (c) comprise a precedence chart for the feasibility study and preliminary design project whose work breakdown structure is represented in Figures 3.1 and 3.2.

The sequence of tasks is dictated by the output-input relations among all the tasks that comprise the project. A task that produces an output which is another task's input must precede the latter task. Thus, the tip or head of an early task leads to the tail of the later task. This is illustrated in Figure 3.3(a), for example, where the tip of the arrow representing "allocate work breakdown structure to staff" leads to the tail of the arrow representing "have staff extend the work breakdown structure." Here the output of the earlier task includes allocated pieces of the work breakdown structure which are essential inputs for the staff to extend their respective parts of the structure.

Sometimes the outputs of several tasks must be available before a subsequent task has all the inputs it needs to start. This is illustrated at point J in Figure 3.3(a), where a cost control plan, a contingency plan, and the extended work breakdown structure must all be available before the feasibility of any of the options can be studied.

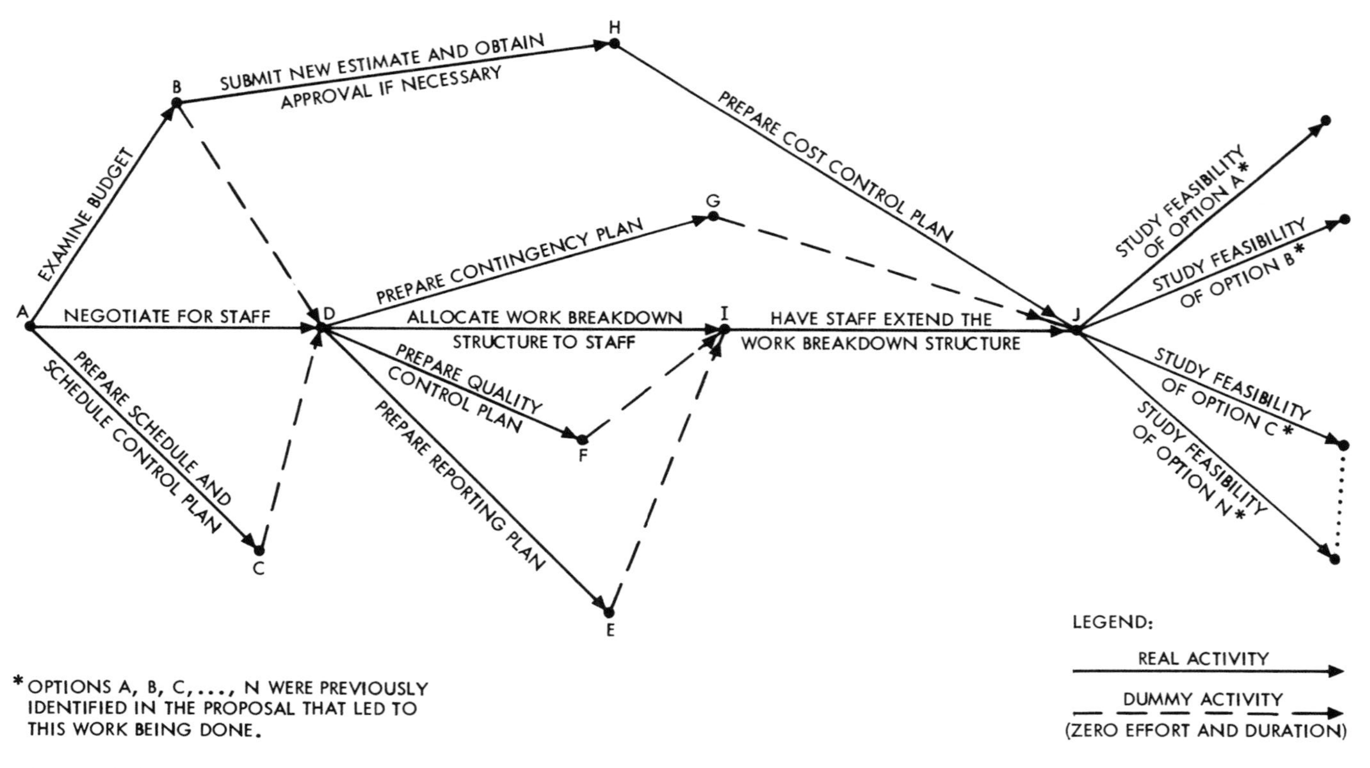

*OPTIONS A, B, C, . . ., N WERE PREVIOUSLY IDENTIFIED IN THE PROPOSAL THAT LED TO THIS WORK BEING DONE.

Figure 3.3(a) The planning segment of a precedence diagram for a feasibility study and preliminary design of a water supply line and supporting facilities.

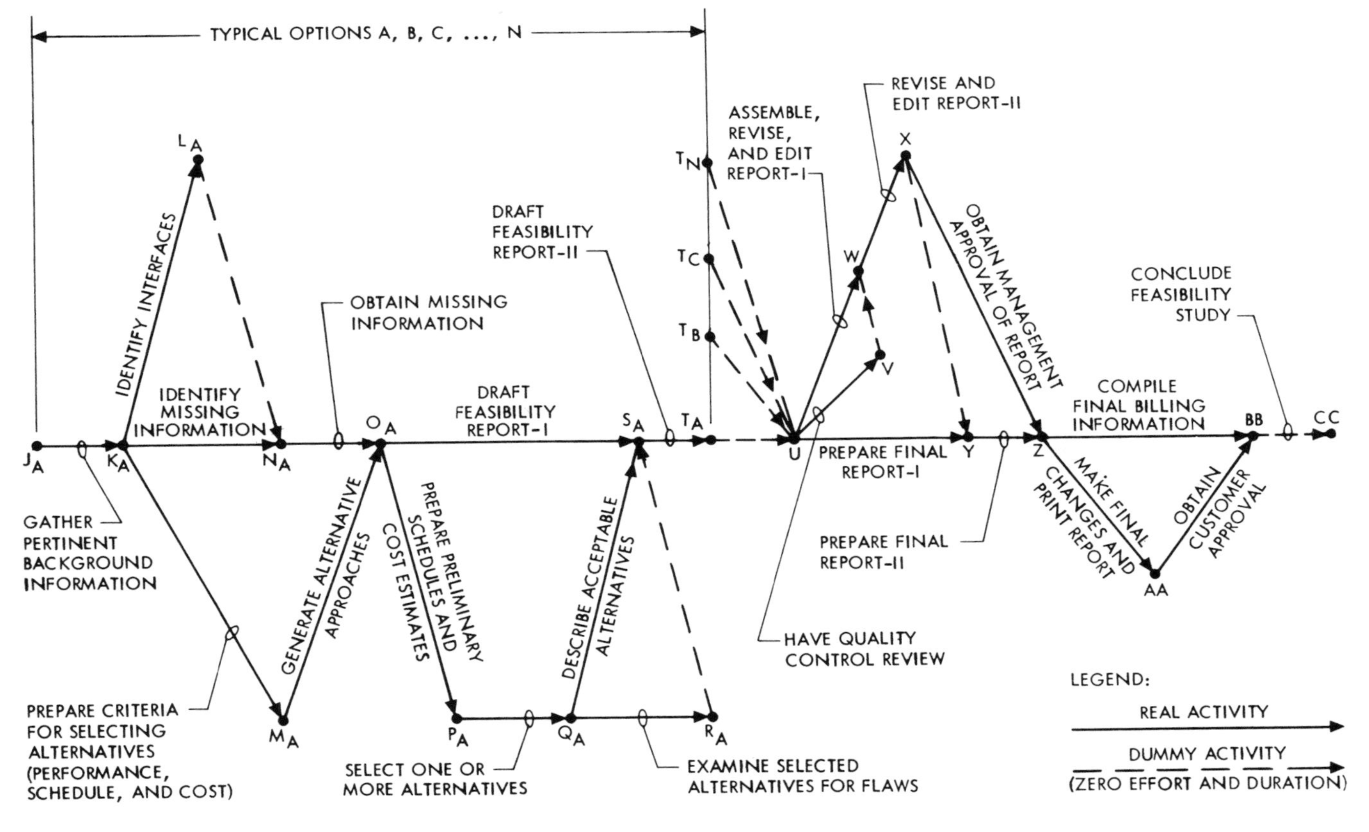

Figure 3.3(b) The feasibility study segment of a precedence diagram for a feasibility study and preliminary design of a water supply line and supporting facilities.

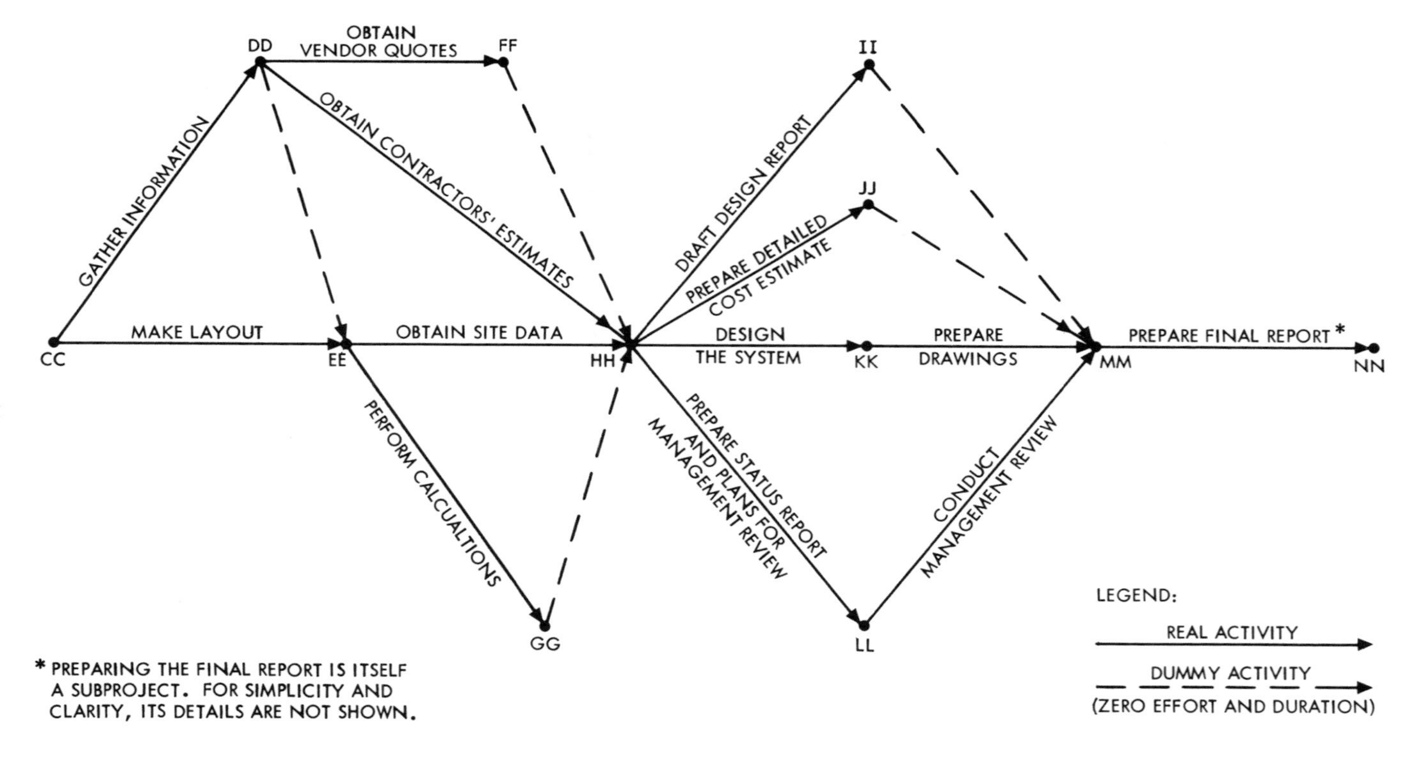

Figure 3.3(c) The preliminary design segment of a precedence diagram for a feasibility study and preliminary design of a water supply line and supporting facilities.

Similarly, the conclusion of a given task can sometimes enable several subsequent tasks. This is illustrated in Figure 3.3(b) at point K. Here, three tasks, "identifying interfaces," "identifying missing information," and "preparing criteria for selecting alternatives," all use as inputs the output of the preceding task, "gathering pertinent background information."

Figures 3.3(a) to (c) also contain some dummy activities, shown by dashed lines. Dummy activities have zero effort and duration and are included to keep scheduling algorithms straight. As discussed in Section VI, scheduling algorithms refer to tasks or activities by the points or nodes at their beginnings (tails) and ends (heads). If two parallel activities had the same nodes for their beginnings and ends, the algorithm could not distinguish between them. To avoid this dilemma, each of the parallel tasks is given a different ending node and then connected to the common ending node by a dummy activity. This is illustrated in Figure 3.3(a) where the task "prepare schedule and schedule control plan" is in parallel with the task "negotiate for staff." The former task is labeled Task A-C and second task is labeled Task A-D. A dummy task, Task C-D, is used to show that both Task A-D and Task A-C must be completed before tasks whose tails lie at point D can be started.

We will return to the subject of precedence charts when we discuss estimating and network techniques of scheduling and budgeting.

IV. SUBPLANS: A CHECKLIST OF MAJOR PLAN ELEMENTS

When one thinks of a plan, it is natural to think mainly in terms of work accomplishment or performance. This is indeed the central part of any plan, but it is seldom the whole plan. The entire list of plan elements is:

1. accomplishment or performance plans
2. manpower and project organization plans
3. quality control plans
4. equipment and material plans
5. work authorization plans
6. cost control plans
7. schedule control plans
8. reporting plans
9. contingency plans

Subplans 2 through 9 are discussed in turn. All that apply should be included in the project plan.

A. Manpower and Organization Plans

The manpower and organization plan tells who will be responsible for each part of the work breakdown structure and what their inter-relationships are. The plan might take the form of a block diagram that shows authority and communication lines (i.e., a conventional organization chart) or the form of a series of role and charter statements, including reporting responsibilities. In practice, a combination of both forms is useful, especially in large, unique projects for which there is no precedent to guide either project staff or other people who will interact with the project.

Often significant negotiations are required to obtain commitments of personnel necessary to fulfill the manpower and organization plan. The project manager can prepare for these negotiations by sketching out a tentative plan, including estimates of the efforts required of each of the key participants.

Preparation of a project manpower and organization plan takes time, which should be provided in the project's overall cost estimates and schedule.

B. Quality Control Plans

The quality control plan contains the scheme for assuring that the project will produce a good product. Thus, it tells who is going to check what, when the checking will be done, and what time and resources are required. It may range from a pair of trained eyes that check the product just before it goes out the door to an elaborate and thorough check and cross check of everything done.

The quality control plan should be well-known to the entire project staff. If one person is to be responsible for assuring quality, that person should prepare the plan, consulting appropriately with those who will do the work that will be checked and with the checkers, as well as with the project manager, so that no one will be surprised.

C. Equipment and Material Plans

The equipment and material plan pertains to the physical resources needed to accomplish the project. It begins with a list of items needed, the dates that they are needed, suggested sources of supply, and the lead-times necessary to obtain the items, including time to obtain price quotations, shipping time, and time to clear customs if applicable. This information together with knowledge of the organization's procurement practices is then used to establish the sequence of activities and the milestones for specifying and ordering every piece of equipment, material, and necessary physical installations, including plumbing and electrical hookups, needed on the project.

Task leaders should prepare equipment and material plans for their respective parts of the overall plan. The project manager should check these plans to be sure that overall project objectives, particularly schedule objectives, will still be met. Serious slippages commonly occur because of inadequate provision for equipment and material leadtimes.

D. Work Authorization Plans

The work authorization plan is the project manager's scheme for approving successive stages of work. It consists of periodic reviews and evaluations to establish the readiness and appropriateness of each task to proceed and a means of authorizing each task leader to proceed when appropriate.

A primary virtue of work authorization plans is that they enable the project manager to revise individual task plans in order to reallocate resources among tasks as overall project needs change. While a project manager can conceivably retrieve resources that were once assigned to task leaders, it is psychologically difficult to do. An easier approach is to release resources incrementally from time to time. This way, the task leaders are less likely to feel abused when they are required to revise their individual plans as a means of optimizing attainment of overall project objectives.

Sometimes the project manager's customer also authorizes work in increments which are then typically called "phases." The customer's motivation is the same, namely to be able to examine the results of early work before setting the course and approving later work.

E. Cost Control Plans

Cost control is based upon cost expectations (i.e., task budgets), cost measurements, budget-measurement comparisons, and revisions of plans and budgets to achieve budget objectives when discrepancies are detected. The cost control plan therefore specifies what budgetary details are needed, what costs will be measured, and what comparisons will be made. It also specifies what techniques will be used to collect and process the information and review it in a timely way so that suitable corrective actions may be taken. These activities require resources and time which must be provided in the overall project plan.

Cost control is discussed in detail in Chapter 4.

F. Schedule Control Plans

Schedule control is based on expectations (i.e., performance schedules), measurements of performance versus time, expectation-measurement comparisons, and revisions of performance plans as needed to achieve performance and schedule objectives. They thus resemble cost control plans in form. The schedule control plan accordingly specifies what performance details will be monitored, when they will be monitored, and by whom. Provisions for these activities must be included in the overall project plan.

Schedule control is discussed in detail in Chapter 4.

G. Reporting Plans

Every project manager will want to know what is going on—all the time. And the customer will want to know on a periodic basis. Thus every project needs a reporting plan that identifies who reports to whom, what is reported, how often reports are made, and how widely the information is distributed.

In order to make overall project reporting as efficient as possible, internal reports (i.e., for those who work on the project) should be coordinated with reports to the customer. Information should be detailed and organized to serve multiple users without recalculation or reformatting whenever possible. Since satisfying customer needs is paramount, their reporting needs should be determined before establishing internal reporting requirements.

As in the case of the other subplans, time and resources must be provided for reporting in the overall project plan. Specific reporting techniques are discussed in Chapter 6.

H. Contingency Plans

Every project needs a contingency plan that tells what will be done if significant planning assumptions turn out not to be true. Planning assumptions may not be true for any of several reasons:

1. inability to predict
2. inability to control
3. lack of information

If a significant planning assumption is false, the project manager may need to negotiate a change in the scope of work, take more (or less) time, or spend more (or less) money in order to accomplish the project.

Neither the project manager nor the customer relishes having to extend the project schedule or budget once they are agreed to. Wise project managers therefore include schedule and financial contingency allowances in plans from the beginning to accommodate untoward consequences of false assumptions. We call these untoward consequences "unexpected needs." They are needs that cannot be specified in advance; otherwise, they would be provided for in the plan.

Examples of unexpected needs are:

1. obtaining customer-furnished information and materials
2. obtaining approvals and permits
3. working in inaccessible locations
4. placing subcontracts and purchase orders
5. obtaining support from within one's own organization
6. following customer procedures
7. developing new techniques
8. replacing sick, vacationing, and reassigned personnel
9. training personnel
10. dealing with tardy vendors, support groups, customs agents, and project staff
11. meeting and talking with others
12. making mistakes

Contingency allowances are different from, separate from, and in addition to the schedule and financial resources determined by good estimating techniques (Section V). Good estimating includes specifying the assumed planning conditions, including the difficulties that are expected. Contingency allowances are then only for the unexpected difficulties.

Contingency allowances should be applied just once in developing a project plan, at the end. They should not be duplicated or "layered" as various plan segments are combined to form larger segments. Otherwise, the accumulated allowances will likely be too generous and make the overall plan too long or too expensive.

The allowances just described are reserves that a project manager and the boss recognize are necessary to protect against damage from incidents such as those listed above. The amounts of these management reserves should not be shared with anyone except the boss. If project staff, subcontractors, or suppliers know of them, they will likely spend them whether they need to or not.

Two other types of reserves are possible in addition to those that the project manager shares with the boss. First, the customer, if there is a customer, may keep schedule and financial reserves when the project is authorized. The project manager may never know what they are, and an unsophisticated customer may not know enough to have them. Do not count on them.

Second, the project manager should have ways of accommodating contingencies that can be called the project manager's reserves. Flexibility or allowances should be provided in budgets, schedules, and even in performance objectives to cope with unexpected needs that cannot be handled with the management reserve. They should be a very private secret and no one should know about them, not even the boss. A smart project manager will find a way to provide for contingencies even if the system does not provide for them.

Sometimes a project is undertaken when everyone recognizes that it is risky. In this case, the project manager should prepare and publish an overall contingency plan that considers and lists danger signs and alternative courses of action for specific eventualities. By publishing the plan, the project staff and even the customer can alert the project manager to early warning signs and thereby help avoid unfavorable events.

Both the work breakdown structure and the precedence chart need to reflect these major elements in addition to showing work accomplishment *per se*, as in Figures 3.1 and 3.3.

V. ESTIMATING TECHNIQUES

The time and expense of performing a project can be estimated in several ways, which are described in this section. No one of the ways is absolutely foolproof, and prudent project managers will use as many as possible and compare their totals. If significant differences are found, then each estimate should be examined to try to account for the differences. Generally this examination will reveal important assumptions or oversights in one or more of the estimates which can then be properly acknowledged, thereby reconciling the differences.

A. Top-Down Estimating

One estimating technique is called "top-down" estimating. In this approach, large blocks of effort are estimated by comparing these blocks with those with which one has had experience. The estimates for the large blocks are then added together to obtain estimates for the entire project.

The top-down approach naturally works best when the blocks in the current project closely resemble those in earlier projects and when accurate records of time and expense exist for the earlier efforts. It works hardly at all when one has no earlier experience to refer to. Its greatest virtues are that it is quickly applied and that it provides for necessary interactions, even if they are obscure.

B. Bottom-Up Estimating

An alternative to "top-down" estimating is the "bottom-up" approach. In this technique, all the little tasks are estimated individually and then summed up to arrive at the total project time and expense.

The "bottom-up" approach depends upon being able to gauge how much time and other resources are needed to perform very elemental tasks. Often one can estimate small increments or elements quite well even when overall efforts cannot be estimated very satisfactorily. This is particularly true when many specialties are involved and no one person has enough breadth to be able to integrate them all together to produce a "top-down" estimate.

A possible problem with "bottom-up" estimating is that the final overall estimate may contain too much contingency allowance for the situation. This arises when each individual estimator provides a minimum contingency allowance to his or her part. In the aggregate, these allowances total more than warranted by the collection of individual elements taken as a whole. Another problem is that this approach can be quite time-consuming, for it requires each individual task to be enumerated and characterized.

C. Choosing an Estimating Approach

"Top down" estimating is most readily performed using a work breakdown structure as the organizing concept, while "bottom up" estimating is best done using a precedence chart as the organizing concept. The reasons are straightforward: A work breakdown structure begins with large blocks of effort which correspond to the units used in "top down" estimating; these blocks do not appear in a precedence chart in a readily recognizable form. A precedence chart, on the other hand, gives all the elemental tasks, including coordinating tasks, that are needed to do a "bottom up" estimate.

D. Standard Costs and Times

An adjunct to either "top-down" or "bottom-up" estimating is the use of standard times and costs. Certain tasks in some organizations, e.g., drilling holes, installing wire, pouring concrete, curing adhesives, taking samples, analyzing samples, writing computer code, etc. are virtually standardized operations. The time and expense involved in such operations are well known and do not vary appreciably from one project to another. Whatever variances exist are randomly distributed and do not depend upon the particular project.

Standard times and costs can be used in preparing estimates without having detailed insight into the tasks themselves. This approach can be used of course at whatever level the standards apply, i.e., either large work blocks or small elemental tasks, but standard information is most likely to exist for individual elemental tasks.

E. Historical Relationships

Other useful adjuncts to "top-down" and "bottom-up" estimating are the historical relationships that may exist among different parts of similar projects. The efforts spent on documentation, quality control, accounting, supervision, and other indirect functions often have rather stable ratios to the expenses for direct labor. Similarly, the costs for piping, instrumentation, insurance, architectural services, etc. may have stable relationships in certain circumstances with the costs of major pieces of equipment such as boilers, heat exchangers, fractionating columns, and so forth.

A special version of the concept of historical relationships is the empirical fact that equipment costs typically vary as a function of capacity or size by a power law of the form

$$\frac{\text{Cost}_2}{\text{Cost}_1} = \left(\frac{\text{Capacity}_2}{\text{Capacity}_1}\right)^n$$

where n is between 0.6 and 0.7. This law is useful in extrapolating data from one or two cases where costs are known or can be easily obtained, e.g., from recent quotations or bids, to cases where they must be estimated. It assumes, of course, that there are neither physical limitations to making the item being estimated nor any step function changes in the design or manufacturing approach between the item whose cost is known and the item whose cost is being estimated.

Manufacturing costs may also vary systematically, but in a complex way, with such parameters as number of separate parts, type of material, amount of material removed by machining, weight or volume of the finished article, and the number of units produced to date. If sufficient historical data exist, statistical techniques can be used to determine parametric relationships of cost versus the variables listed. These relationships can then be used to predict the cost of new items that are made essentially the same way as those for which there are data.

F. Retaining Estimates

Astute project managers, and indeed astute organizations, recognize the value of saving both cost estimates and actual cost data for future reference. In the final analysis, there is no better basis for estimating the next project than the data developed on related or similar previous projects. Also, as will be seen in Chapter 4 on project control, the estimates used to prepare the project schedule and budget will provide the basis for controlling the project. It pays, therefore, for the project manager to take care over the estimating process and attempt to produce estimates which are as meaningful and detailed as possible.

VI. SCHEDULING TECHNIQUES

The essence of project scheduling is to apply time estimates to the precedence chart and then to make various adjustments, which are discussed later, so that the entire project can be accomplished within the time and budget allowed.

If "bottom up" estimates have been prepared, then estimates can be applied to the individual tasks in the precedence chart. This is illustrated in Figure 3.4, using Option A of the project represented in Figure 3.3(b) as an example. If only "top down" estimates have been prepared, then the estimates can be applied to segments of the chart each of which includes many individual tasks. This is illustrated in Figure 3.5, using Option B of the same project as the example.

When "top down" estimates are used, one or more coordinating tasks may be necessary but overlooked in the estimated segments. The time required for these tasks must also be estimated and applied to the precedence chart.

Once the time estimates have been applied to the precedence chart, the longest sequence from project beginning to project completion can be determined. The procedure can be illustrated by referring to Figure 3.4 and the following calculations.

First, the earliest possible starting time and earliest possible completion time are calculated for each task. No task can start sooner than the completion of all the tasks that must precede it. When two or more tasks immediately precede a given task in parallel, both (or all) of them must be completed before the given task can begin. Thus, the completion of the last of the parallel tasks governs the start of the subsequent task. Earliest possible starting and completion times are shown in columns 3 and 4, respectively, of Table 3.1, which relates to Figure 3.4.

Note that the end of the project is indicated by the beginning of an imaginary task of zero duration, called "conclude feasibility study." "Conclude feasibility study" is added to the real tasks in order to maintain the system or algorithm for

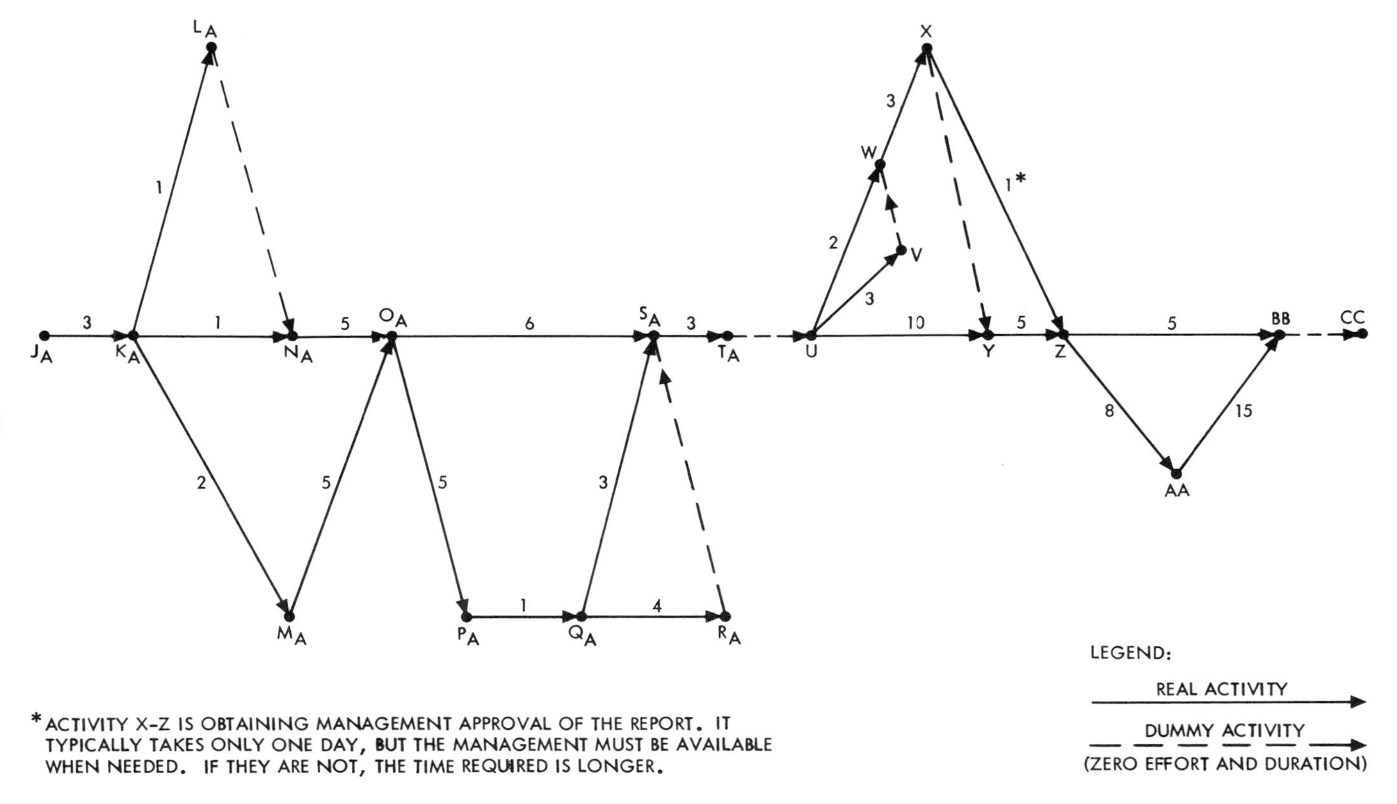

Figure 3.4 Time estimates for Option A of the precedence diagram shown in Figure 3.3(b).

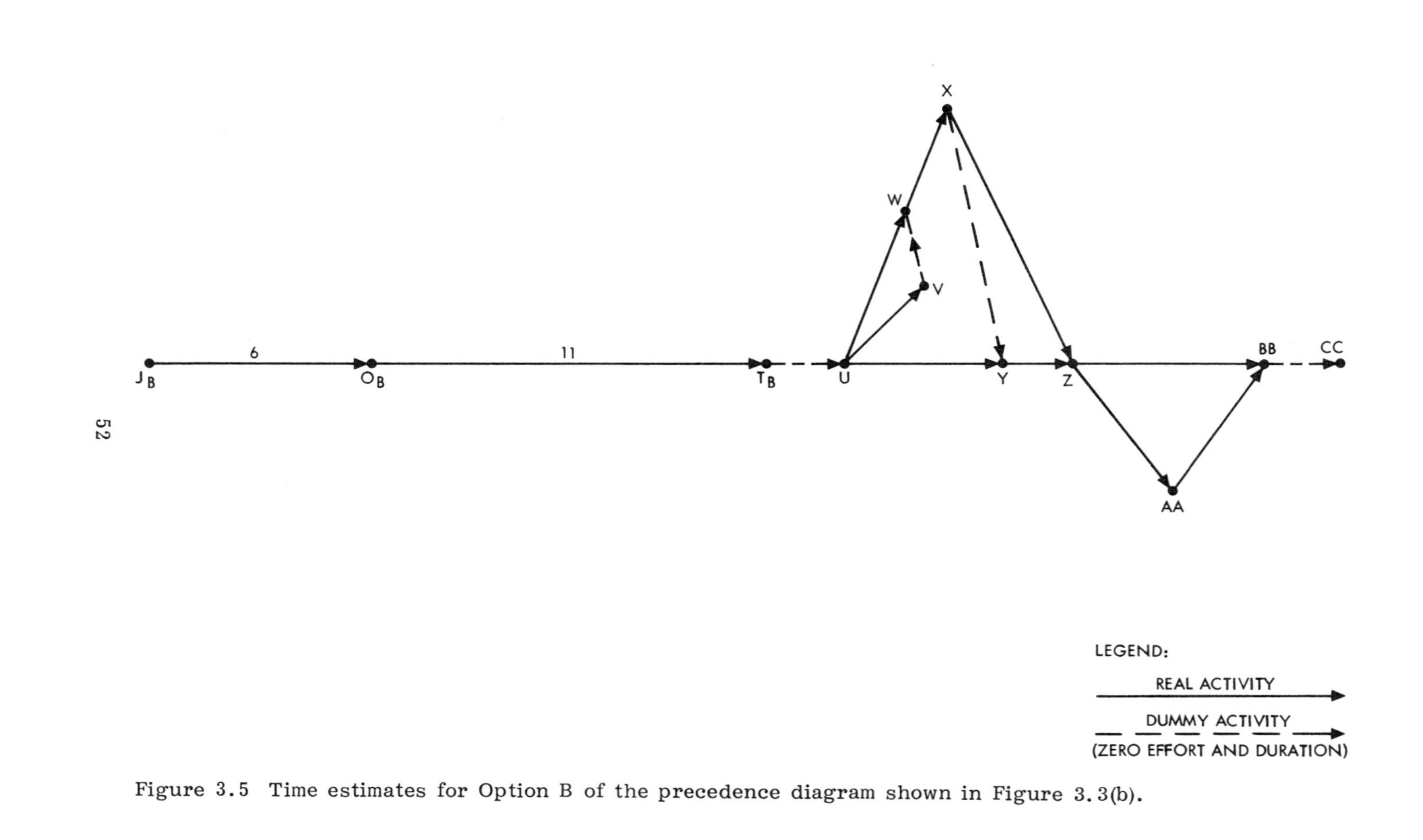

Figure 3.5 Time estimates for Option B of the precedence diagram shown in Figure 3.3(b).

calculating dates. Its beginning and end both coincide with the later of the two completion times for tasks Z-BB and AA-BB. It thus removes any ambiguity there might be about when the real work is in fact concluded.

The next step is to calculate the latest permissible starting times for each task. These times are the latest times that the tasks can be started without causing any subsequent task to be so late that it delays the task called "conclude feasibility study." The latest permissible starting times are calculated, of course, by starting at the "conclude feasibility study" task and working backwards. These times and the corresponding latest permissible completion times are shown in columns 5 and 6, respectively, of Table 3.1.

The difference between the earliest possible and latest permissible starting times for each task is called the "slack" for the task. Slack refers to the amount of delay that a task could have, relative to its earliest possible starting time, without causing a delay to the overall project. Tasks which have zero slack are said to be "critical." Critical tasks cannot be delayed without causing the project to be delayed overall. The sequence of critical tasks, which there will certainly be, is called the critical path for the project. The critical path for the example in Figures 3.3(b) and 3.4 is shown in Figure 3.6. Figure 3.7 is a time-scaled version of this critical path.

If the overall project duration, i.e., the duration of the critical path is within acceptable bounds, no adjustments in the project schedule are needed. In this case, the project manager must simply make sure that (1) no critical task slips schedule to the point that it causes an unacceptable slip in the overall project and (2) no task which is not critical slips so much that it becomes critical. If the overall project duration as just estimated is unacceptably long, however, then the project manager must find ways to shorten the project duration. He or she may also wish to shorten one or more segments of the project if there is an expensive item whose cost could be decreased by using it more intensively or minimizing delays between successive uses.

One approach to shortening the project's duration is to "double up" on one or more of the tasks that lie on the critical path so that the total project can be done in less time. In some cases this can be done without extra cost or at least with no more cost than that saved by doubling up. In others it may require extra funds, e.g., to pay for overtime or for extra coordination required when a task is subdivided and

TABLE 3.1
Critical Path Calculations for Option A of the Precedence Diagram Shown in Figure 3.3(b)

Task	Duration	Earliest Possible Start Time	Earliest Possible Completion Time	Latest Permissible Start Time	Latest Permissible Completion Time	Slack	Critical Path
J_A-K_A	3	0	3	0	3	0	Yes
K_A-L_A	1	3	4	4	5	1	No
K_A-M_A	2	3	5	3	5	0	Yes
K_A-N_A	1	3	4	4	5	1	No
L_A-N_A	0	4	4	5	5	1	No
M_A-O_A	5	5	10	5	10	0	Yes
N_A-O_A	5	4	9	5	10	1	No
O_A-P_A	5	10	15	10	15	0	Yes
O_A-S_A	6	10	16	14	20	4	No
P_A-Q_A	1	15	16	15	16	0	Yes

Q_A-R_A	4	16	20	16	20	0	Yes
Q_A-S_A	3	16	19	17	20	1	No
R_A-S_A	0	20	20	20	20	0	Yes
S_A-T_A	3	20	23	20	23	0	Yes
T_A-U	0	23	23	23	23	0	Yes
U-V	3	23	26	27	30	4	No
U-W	2	23	25	28	30	5	No
U-Y	10	23	33	23	33	0	Yes
V-W	0	26	26	30	30	4	No
W-X	3	26	29	30	33	4	No
X-Y	0	29	29	33	33	4	No
X-Z	1	29	30	37	38	8	No
Y-Z	5	33	38	33	38	0	Yes
Z-AA	8	38	46	38	46	0	Yes
Z-BB	5	38	43	56	61	18	No
AA-BB	15	46	61	46	61	0	Yes
BB-CC	0	61	61	61	61	0	Yes

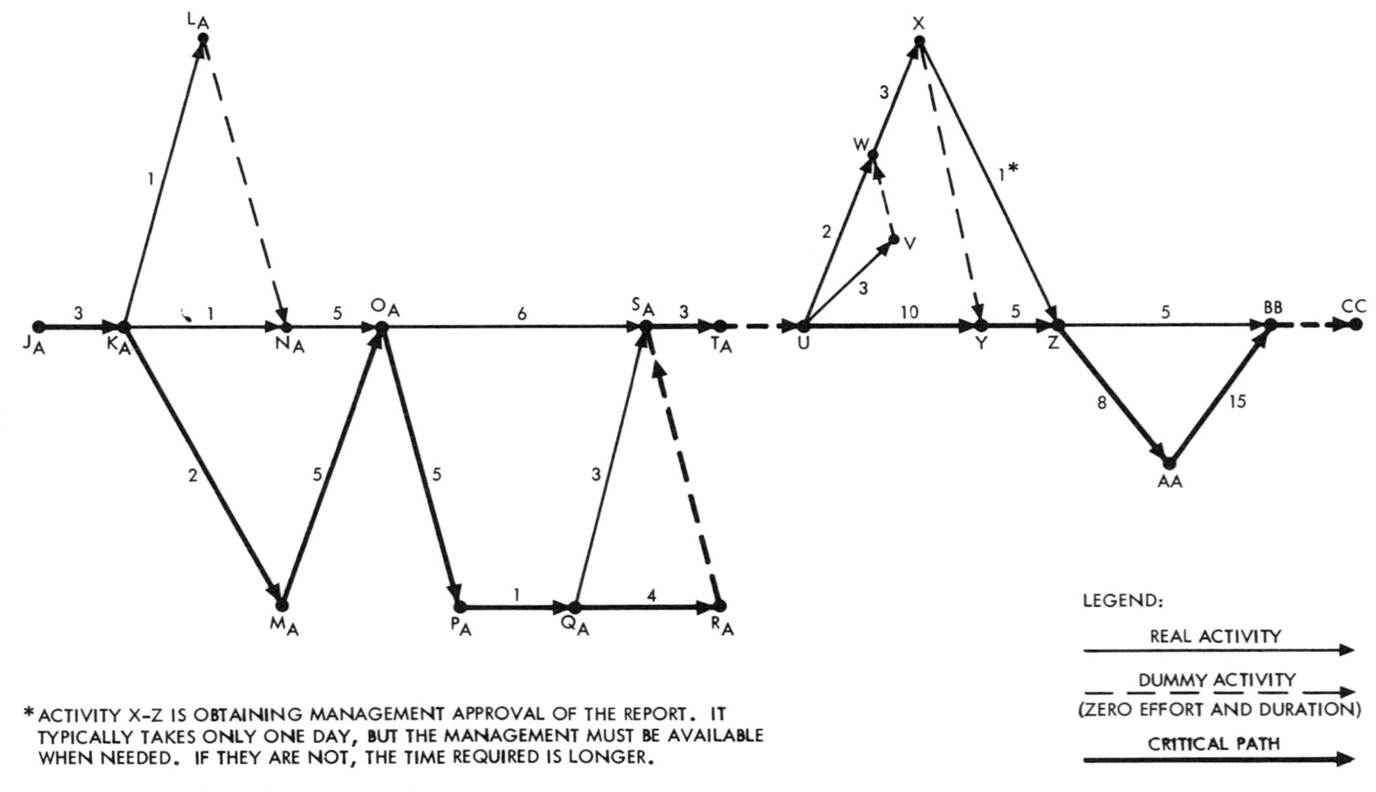

Figure 3.6 The critical path for the precedence diagram shown in Figures 3.3(b) and 3.4.

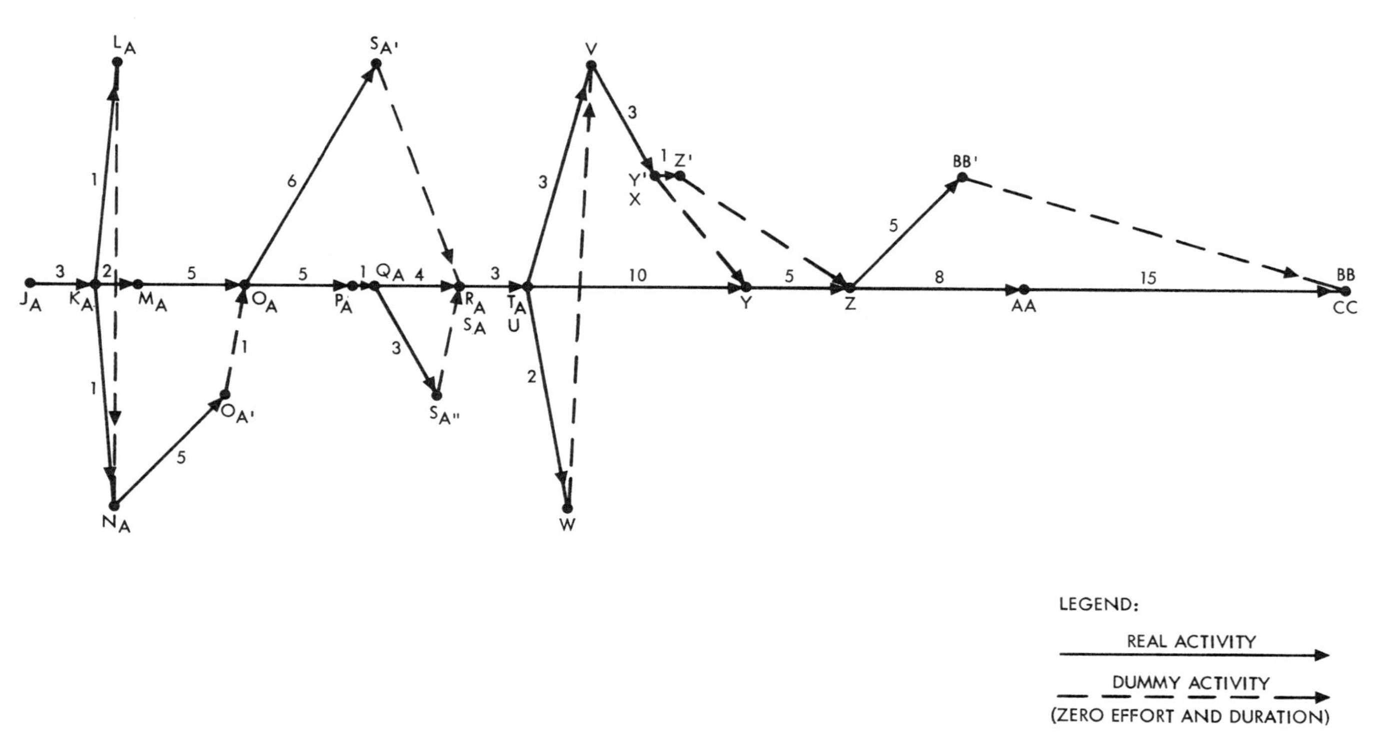

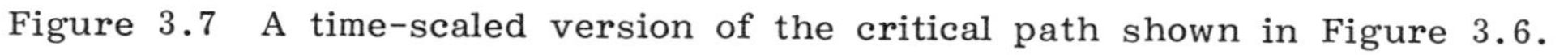

Figure 3.7 A time-scaled version of the critical path shown in Figure 3.6.

assigned to different people. If extra costs are involved, they must be agreed to by the customer before they are incurred. In still other cases, it may not be possible to double up (viz. two elephants each pregnant for 10½ months is not equivalent to one elephant pregnant for 21 months!). Another approach is to negotiate a reduction in the total project scope of work so that it can be accomplished within an acceptable time. And, of course, it is sometimes possible to negotiate an extension in the time allowed to complete the project.

It is possible that a suitable combination of these three approaches is best overall. The preferred mix will depend upon the customer's or boss's priorities, and the project manager should be prepared to discuss the pros and cons of various ways to define and schedule the project.

The above description of scheduling uses what might be called "nominal" time estimates for each task. Some refinements are possible if it is recognized that the estimates might not be exact. If, for example, the probability distribution of possible times is known or can be estimated for each task, then their most likely times can be calculated and used. Or the times that give a particular level of confidence, say 95%, that the project will not exceed the projected overall duration can be used.

A particularly useful refinement is to use a weighted average that takes into account the fact that the actual time may be different from the nominal but is unlikely to exceed some maximum value for the task and is unlikely (or unable) to be less than some minimum value. A common weighting function for the time estimated for task i, $t_{i_{est}}$, is

$$t_{i_{est}} = \frac{t_{i_{min}} + 4t_{i_{nom}} + t_{i_{max}}}{6}$$

where

$t_{i_{min}}$ is the minimum possible duration of task i,

$t_{i_{nom}}$ is the nominal duration of task i,

and

$t_{i_{max}}$ is the maximum likely duration of task i.

A further refinement in applying this scheduling approach is to enlist a computer to perform the calculations, which are admittedly tedious and subject to error. Many commercial computing services and software packages have canned scheduling programs that will determine slacks, critical paths, and overall project duration from task time estimates and precedence relations. These programs will even graph the schedule.

A computer approach is very useful if one wants to evaluate the effect of "doubling up" or "stretching out" an activity or of alternative precedence relations. An example of an alternative is to do some tasks in parallel that would logically be done in series. If this approach is used, some tasks may need to be redone if they are based on assumptions that are found to be false upon completing a parallel activity. In this case, provisions must be made in the schedule to allow for the revisions.

Another reason for using a computer in scheduling is that it facilitates keeping the schedule current when better, i.e., more accurate, estimates become available. Typically, a project manager can improve the quality of the time estimates for later tasks as project work progresses. By using a computer to keep track of the schedule, the project manager can revise it easily as better time estimates are made for upcoming tasks.

The scheduling approach just described is variously known as the critical path method (CPM) or as program evaluation and review technique (PERT). The particular representation used here is called "activity on arrow," because the activities are in fact represented by the arrows or lines of the drawing. An alternative representation is "activity on node," where the activities are just shown by points connected by lines. In this latter case, there is no time scale implied by the horizontal axis, just sequencing relationships. Many computer printouts use the "activity on node" representation because it saves paper.

Before leaving the subject of scheduling, one more comment is needed, on contingency allowances. An overall schedule contingency pertains to the critical path. This is true because non-critical tasks have slack to cope with any slippage they may incur (see Chapter 4). Of course, if a non-critical task consumes its available slack, then it becomes critical and will need to draw upon the project's schedule contingency if it slips further.

Note also that a budgetary contingency allowance is probably necessary in order to use both slack and the schedule contingency. That is, if work slips schedule, there will be some on-going overhead costs to be paid during the extra time that is used to perform the work, even if the work itself is not more costly. Moreover, work whose schedule slips may also cost more *per se*, whether in critical or non-critical tasks. If the slippage is due to having to redo some of the work, then there is clearly the extra expense of redoing it. And if work slips enough to crowd an important completion date, extra efforts are often needed to expedite it or to "double up" so that part of the slippage can be recouped.

VII. REPRESENTING THE PLAN

A fully developed project plan contains so much detail that it is difficult for many people to comprehend its most important features. Accordingly, a project manager may find it expedient to summarize or capsulize it when talking with individuals who are not interested in its intricacies.

Descriptions of the project's main objectives, main approaches, main management techniques, etc. can be reduced to just a few pages of narrative. If the description exceeds four or five pages, most people who should read it probably will not do so.

Budget summaries can be prepared by major objectives, major tasks, or major blocks of time. Supporting elements, e.g., quality control, can be included with the major elements or listed separately, depending upon the need for simplicity or explicitness.

Schedule information can be handled by a simplified CPM or PERT network chart that treats many related tasks as a single work unit, as in Figure 3.8. Alternatively, schedule information can be represented in a bar chart, as in Figure 3.9. This figure gives the same schedule information as Figure 3.4. While it does show time phasing of the tasks, it does not reveal the essential precedence relationships that connect them.

A special version of the bar chart is the Gantt chart, after the man who developed it. In a Gantt chart, each task

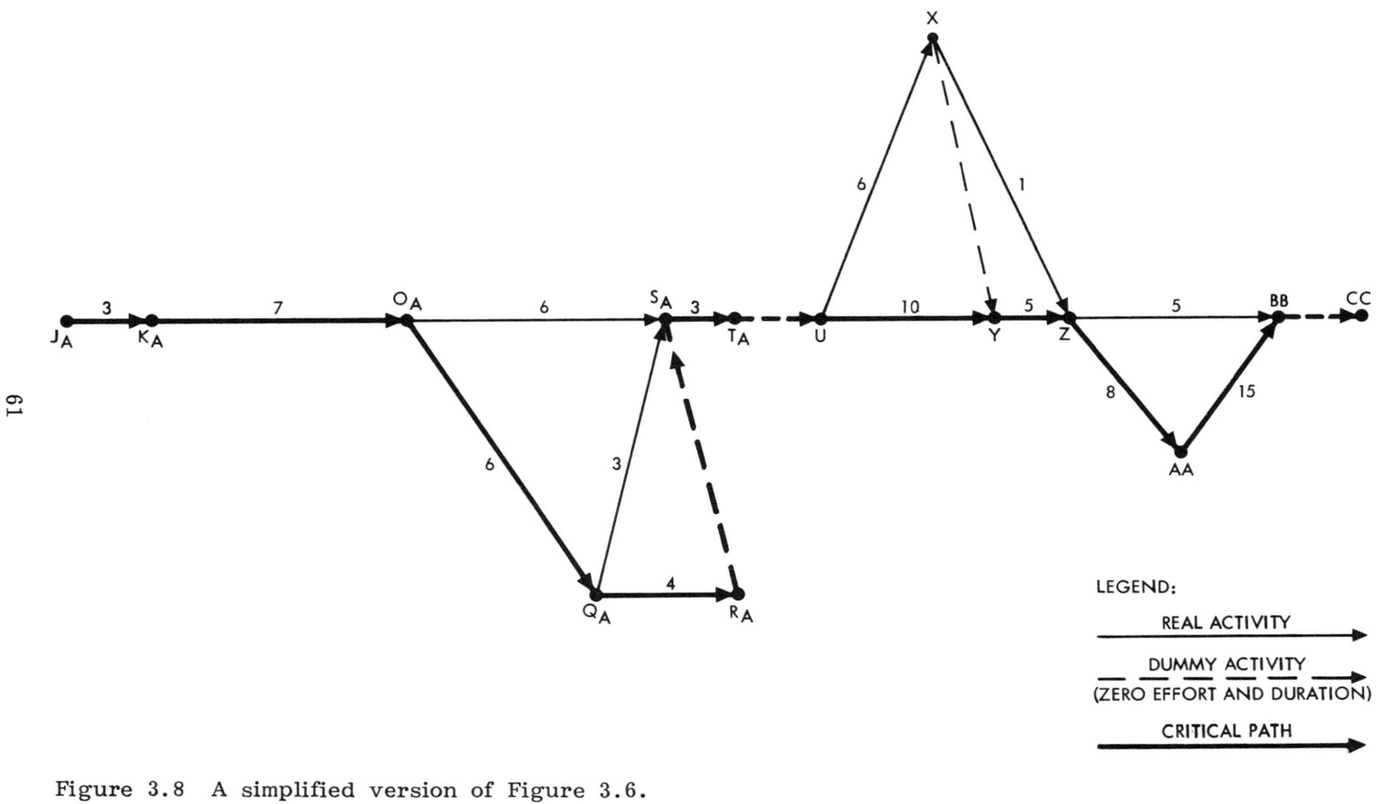

Figure 3.8 A simplified version of Figure 3.6.

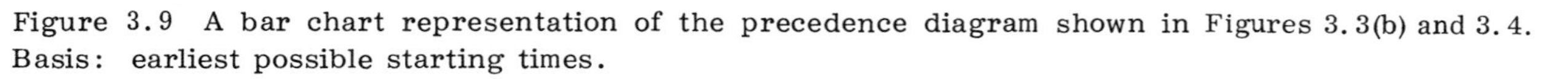

Figure 3.9 A bar chart representation of the precedence diagram shown in Figures 3.3(b) and 3.4. Basis: earliest possible starting times.

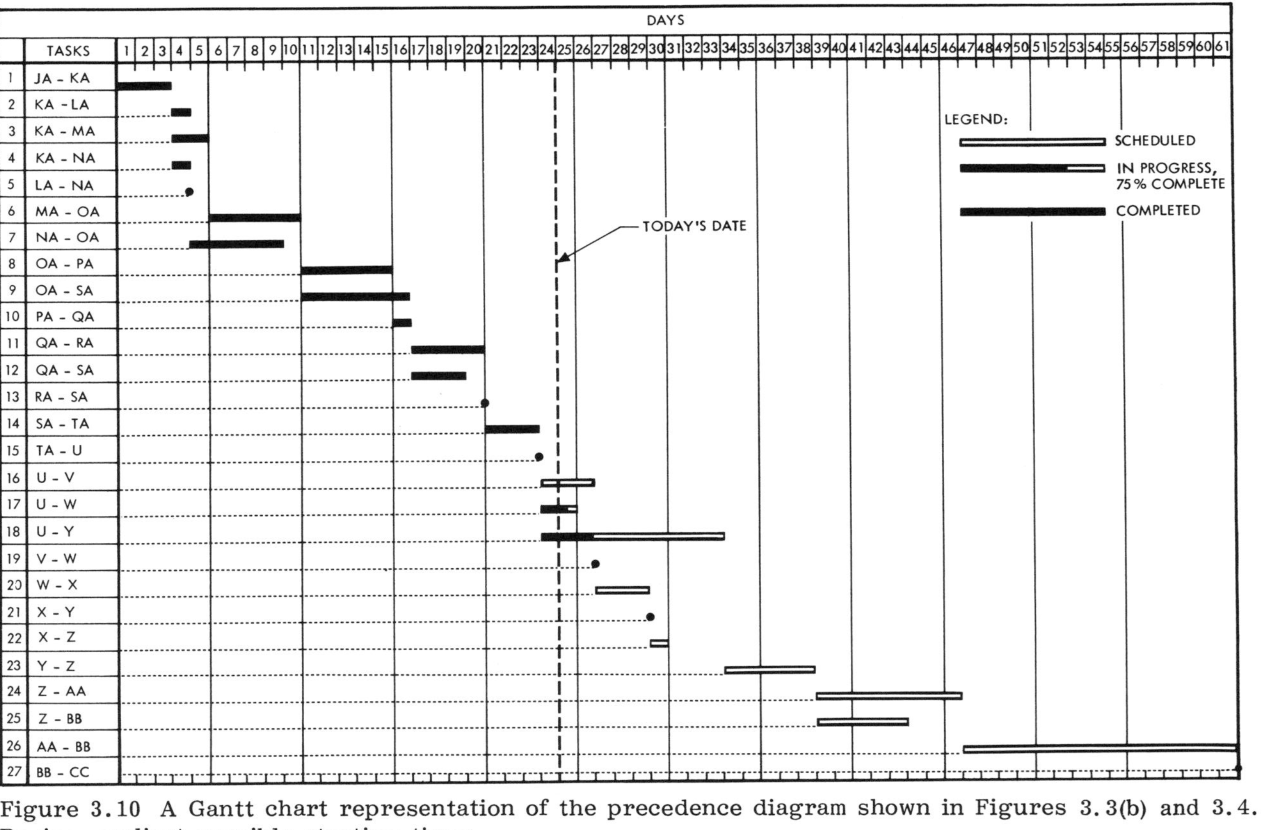

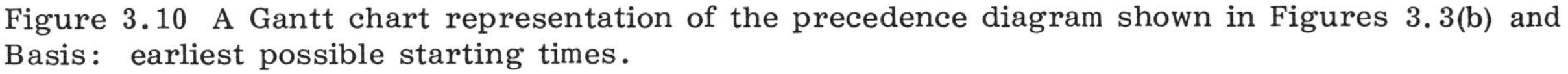
Figure 3.10 A Gantt chart representation of the precedence diagram shown in Figures 3.3(b) and 3.4. Basis: earliest possible starting times.

bar is shaded to show its percentage completion as of the date of the chart, as in Figure 3.10. *If* it is assumed that the work of each task should proceed at a uniform rate during the task, then the percentage completion data can be directly compared against the date of the chart to determine which tasks are ahead, behind, and on schedule. However, many tasks do not proceed at a uniform rate, so such comparisons may not be meaningful.

4

Control Techniques

I. PRINCIPLES OF CONTROL

Project objectives are rarely met and met on time and within budget unless the project is controlled. By "controlled," we mean "monitored and managed in such a way that deviations from plan are detected and corrected in time for the objectives indeed to be met and met on time and within budget." Thus control implies:

1. expectations of what should happen
2. measurements of what is happening
3. comparisons between expectations and what is happening (or has recently happened)
4. timely corrective actions designed to meet the objectives, schedule, or budget

If any one of these four steps is missing, the project cannot be controlled.

This simple yet powerful concept of control contains some important corollaries. First, the expectation or plan must be expressed in terms that are also suitable for measurement; otherwise, it will not be possible to compare measured results with expectations. Second, what is measured should correspond to elements in the plan; otherwise, necessary comparisons cannot be made even if the plan is expressed in measurable terms. And third, the plan and the corresponding measurements must be made at intervals close enough together that when a deviation is detected there are still enough time, budget, and other needed resources available to correct the situation.

It follows from this last corollary that the less contingency allowance (or slack) available, the more closely a project must be monitored for control purposes. A project manager does not want to find that needed corrections (including perhaps completely redoing a segment of the work) cannot be made within the time or resources remaining. Thus, when contingency allowances are minimal, the project plan and the monitoring program must provide information on relatively small increments. Otherwise, the deviations will not necessarily be small enough to be corrected without exceeding available contingency allowances. Conversely, when contingency allowances are generous, the plan and monitoring program can provide information infrequently and on relatively large increments of work, because the allowances can accommodate even relatively large corrections.

The fact that the level of monitoring permissible or required varies as a function of the contingency allowance available leads to the notion that there may be an optimum mix of monitoring and contingency allowance. If an infinite contingency allowance is provided, monitoring can be reduced to virtually nothing: The project manager can wait until the end to see if everything worked out. If it did not, then the work can be redone (for in this extreme case the contingency allowance is infinite). This approach is very cheap in terms of monitoring and infinitely expensive in terms of contingency allowance. While the contingency might not be used, allowing for it in case it is needed can result in scheduling and pricing the project beyond the customer's ability to wait or pay.

At the other extreme, the monitoring program can be both very intensive and extensive, so that incipient deviations from plan are detected and corrected almost instantaneously; this approach uses virtually no contingency allowance. However, while quite cheap in terms of contingency allowance, it is very costly in terms of monitoring. Again, the costly element, i.e., monitoring in this case, may exceed the customer's budget.

An intermediate blend of monitoring and contingency allowance avoids the two very costly extremes and can be suitably effective. This effect is depicted in Figure 4.1, where the costs of monitoring, of contingency allowance, and of their total are shown for a given level of confidence that project objectives, schedule, and budget will be met. This figure shows a minimum in the total cost. At this minimum, the sum is less than the cost in either case where one of the components is zero. That is, a blend of both monitoring and contingency allowance is more cost-effective than either one or the other alone.

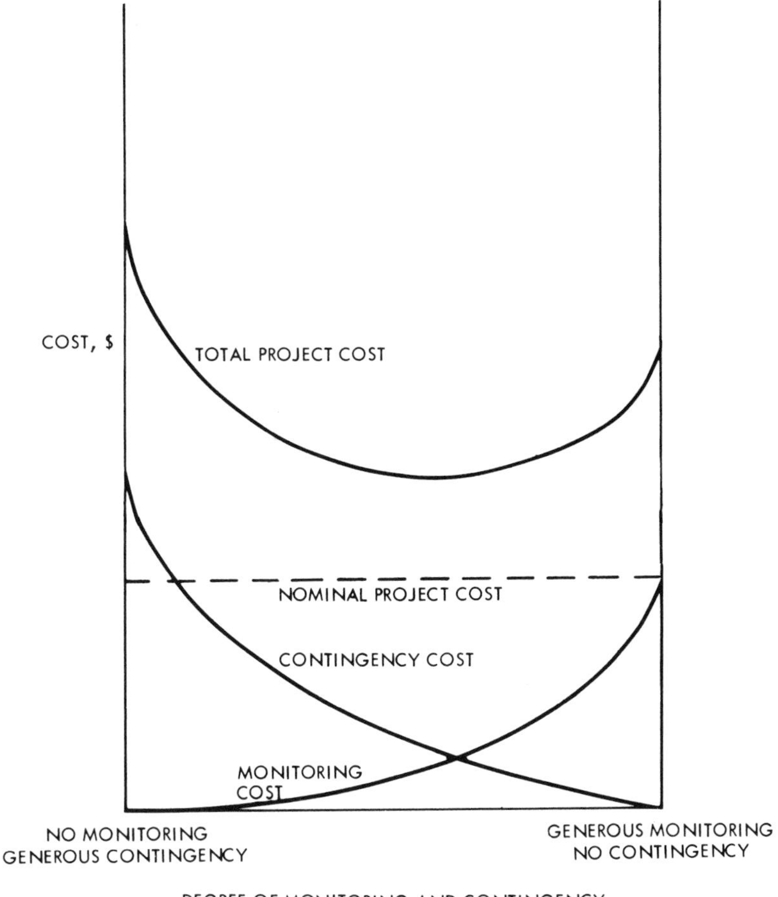

Figure 4.1 Project cost vs degree of monitoring and contingency for a given level of confidence (probability) that project objectives will be met. (From Ruskin, A. M., 1981.)

Figure 4.2 shows schematically the relation of total costs of control for different levels of confidence that the project objectives, schedule, and budget will be met. Curve A in this figure is the same as the total cost curve in Figure 4.1, while curve B represents total costs for a higher level of confidence, and curve C represents total costs for a lower level of confidence.

The curves in Figure 4.2 refer only to levels of confidence, or probabilities, that project objectives, schedules, and budgets will be met. In other words, good project control does not guarantee project success; it can only raise the likelihood of success to a very high probability. Even then, an occasional project, appropriately controlled, will not fully meet an objective, schedule, or budget. Nothing in project management is so certain as to prevent all such occurrences.

The general principles just described can be reduced to specific procedures for task or performance control, schedule control, and budget control. They are discussed in the following sections.

II. TASK CONTROL

Task control consists of assuring that the work itself is accomplished according to plan (without regard to schedule or budget, which are handled separately).

A. Detailed Functional Objectives

Task control is based first of all on detailed functional objectives for the work items in the work breakdown structure (WBS) described in Chapter 3. The objectives need to be written in such a way that their accomplishment can be ascertained unambiguously. Work items or work packages in the WBS should be monitored in increments which are no greater than their associated contingency allowances or slacks, and preferably much less. Then, if a particular piece of work is found deficient, it can be redone within the available allowance and not thwart the attainment of the project's overall objectives, schedule, and budget. By monitoring in increments that are relatively small compared to the allowance for the segment, correction of a single deficiency

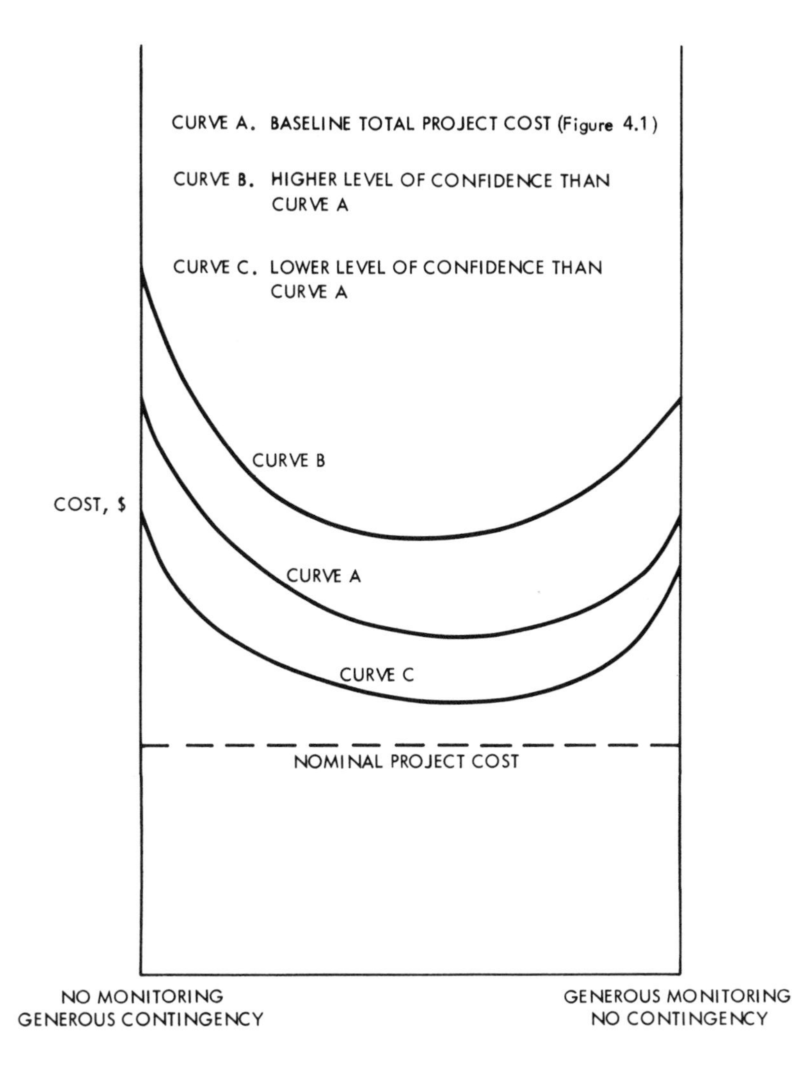

Figure 4.2 Total project cost vs degree of monitoring and contingency for different levels of confidence (probability) that project objectives will be met. (From Ruskin, A. M., 1981.)

will not consume a major part of the allowance. This will leave some allowance for correcting other increments that might later be found deficient.

B. Quality Reviews

Quality control reviews and inspections are an important aspect of task control and go hand-in-hand with detailed functional objectives. Each work increment, no matter how simple, needs to be checked *at some level* to assure that it is satisfactory. Project managers dare not find that a trivial item, such as late delivery, wrong specifications, inadequate access, incorrect size, faulty instrument, lack of suitable personnel, and so forth, is precluding a successful project. And, certainly, no major task can go unchecked. Thus, project managers must arrange for timely and appropriate reviews or inspections to confirm either that the work is being done according to plan or that a deficiency exists which must either be corrected or accommodated by some other change in the plan.

C. Work Orders

Another element in task control is the use of work orders to authorize work increments or packages. By using work orders, the project manager can force personnel working on related parts of the project to coordinate their efforts: They may not proceed until authorized and the authorization will be given only when the project manager is satisfied that their respective efforts properly reflect the interactions. This approach minimizes the chance of having to redo a portion of the work because of overlooking interactions of related activities.

Work orders also preserve flexibility for major revisions which may be necessary if contingency allowances are fully consumed. When an original contingency allowance is used up, the only way the project manager can provide for still other possible deficiencies and unforeseen events is to rescope some or all of the remaining work. While it is theoretically possible to cancel work already authorized, it is time-consuming and perhaps costly to do so. On the other hand, if most of the future work is yet to be authorized, some of it can easily be "cancelled," simply by not authorizing it. Thus, the use of work orders gives project managers an opportunity to rescope future work in light of the total situation and thereby better meet their overall goals.

III. SCHEDULE CONTROL

Schedule control consists of assuring that the work is accomplished according to the planned timetable. Generally, there is little concern if work is accomplished early, so attention is usually focused on preventing schedule slippage. However, if premature accomplishment would result in cash flow problems or excessive interest charges, project managers should keep major activities from occurring until they are needed.

A. Causes of Schedule Slippage

Schedule slippage is commonplace and should be a major concern of project managers. Slippage occurs one day at a time and project managers need to be ever-vigilant to keep slippage from accumulating to an unacceptable level. Since slippage is not a problem where there is sufficient schedule slack but is a problem where there is either insufficient or no slack, project managers cannot wait until slippage has become conspicuous. Rather, they need to have early warnings of potential schedule slips. These can be gleaned from the behavior of project staff by understanding the causes of slippage.

Slippage can be caused by complacency or lack of interest, lack of credibility, incorrect or missing information, lack of understanding, incompetence, and conditions beyond one's control, such as too much work to do. Project managers need to be on the alert to detect the existence of any of these factors in order to nip them in the bud before they result in schedule slips. (The existence of one or more of these factors does not guarantee that a schedule will be slipped. It merely provides the basis for a schedule slip and at the same time provides project managers with early signs of possible slips.)

B. Prevention of Schedule Slippage

Schedule slips are not inevitable; in fact there is much project managers can do to limit them. They can, for example, subdivide tasks and assign responsibilities, establish check points and obtain timely and meaningful reports, act on difficulties, and finally show that it matters. Together, these steps will counter complacency, lack of interest, and lack of credibility. They will also provide opportunities to correct erroneous information and supply missing information, to build understanding, and to detect incompetence and conditions beyond the project staff member's control. If done seriously and with reasonable thoroughness, project managers stand an excellent chance of meeting their target schedules, assuming of course that they are realistic. There is nothing project managers can do to meet truly impossible schedules!

C. Impossible Schedules

Project managers should not agree to schedules they know are impossible to meet. If, however, they agree to schedules which they originally thought possible to meet but later discover are not possible, then they need to renegotiate their commitments or schedules as soon as they detect the difficulties. To do any less is to be unfair both to their customers and their project staffs.

IV. COST CONTROL

Cost control consists of assuring that work elements are accomplished within their respective budgets. Because of their differing characteristics, it is useful to have three separate budgets for each work element: a budget for direct labor, a budget for support services, and a budget for purchased services, materials, and equipment.

A. Budgets

1. Direct Labor Budgets

Expenditures for direct labor occur in relatively small units, perhaps as small as a fraction of an hour, and may be spread over a large number of people who are directly or indirectly responsible to the project manager. The expenditures are typically not coordinated at the level where they are incurred, so the project manager must establish individual budgets for the various work elements in order to be sure that their total does not exceed the total budget available for this work. Otherwise, "minor overages" for each individual element will accumulate to a major overage for the project manager, one that exceeds any reasonable contingency allowance there might be.

2. Support Services Budgets

Expenditures for support services need to be handled separately because these functions are typically provided on a "time and materials" basis with no budgetary discipline. Support service groups will generally do whatever is asked of them and will charge accordingly, making no attempt to absorb costs due to errors of estimation or performance. Moreover, their work tends to be done in large lumps, so that their overages are also lumpy and come without much warning. Thus, the opportunity to revise their efforts in order to accommodate their mistakes and deficiencies is limited. However, work assigned to support services tends to be more familiar than exotic, and it can be characterized and scoped quite well if the effort is made to do so. Accordingly, project managers should insist on obtaining detailed budgetary estimates for work to be done by support groups so that they can be scrutinized for completeness and accuracy. Since there is little chance to adjust budgetary allocations for support services later, they must be prepared with care at the beginning.

3. Budgets for Purchased Items

Expenditures for purchased services, materials, and equipment may be made on a basis ranging from "time and materials" to "firm fixed price." The former need to be budgeted with the same care as support services, for the same reasons. The latter may require less detailed information for budgeting purposes, but they have other problems.

For example, suppliers may fail to deliver because they cannot perform for the price agreed to. Or suppliers may try to renegotiate the price for the same reason. Or unacceptable work or material may be supplied to meet the agreed price.

Suppliers might not provide timely notification of any difficulties, even when required by contract to do so. Also, contracts to suppliers are difficult to adjust unilaterally, and commitments to suppliers are thus harder to break than internal arrangements.

For all these reasons, budgets for services, materials, and equipment to be procured from suppliers should be identified separately and carefully prepared.

B. *Expenditure Reports*

Once budgets have been suitably prepared for the various components of the project, expenditures should be reported against them periodically. In the case of purchased services, materials, and equipment, the figures reported should be funds committed rather than invoices received. To wait for an invoice is to harbor a false impression about how much money is left to accomplish the remaining work.

C. Expenditure Audits

Expenditures should be audited to verify (1) that they refer to legitimate charges to the project, and (2) that the work to which they refer was in fact done. Strange as it may seem, project budgets are occasionally charged for work that should have been charged to another project and for work which is yet to be done. The first type of false charge may be simply an error. Both types could be the result of chicanery on the part of an unscrupulous staff member. A project manager cannot afford to have the project budget pilfered by either type of false charge. Expenditure auditing will help curb these abuses.

D. Comparing Expenditures Against Budgets

After expenditures have been checked to verify that they are legitimate, they should be compared to budgets and to the work accomplished. The comparisons can take any of several forms. One possibility is to compare expenditures for each period to the budget for the period. This approach indicates how well the project is going currently, but it says little about how well it is going overall. To overcome this shortcoming, accumulated expenditures to date can be compared to the accumulated budget to date. This latter approach is particularly useful to tell if the project is in more or less budgetary difficulty than it was in the preceding period. It is also useful in determining the extent to which an accumulated overage is within the budgetary contingency allowance.

Graphs are often helpful when comparing accumulated expenditures against the accumulated budget. Figure 4.3 shows a typical accumulated budget curve on a linear scale, and

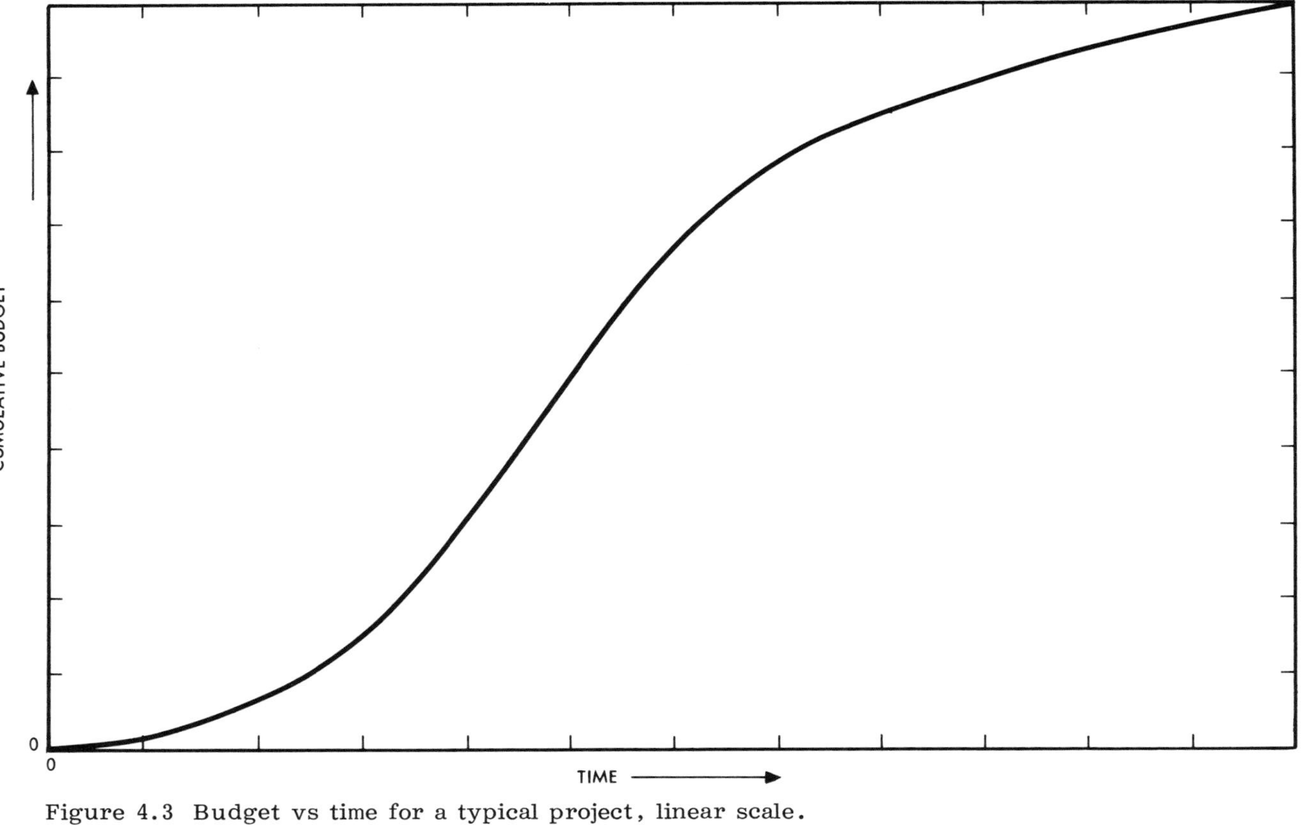

Figure 4.3 Budget vs time for a typical project, linear scale.

Figure 4.4 shows the same information on a semilogarithmic scale. Each scale has its advantages as a control device.

The linear scale provides the same precision at the end of the project as the beginning, that is, very good precision, and it is at the end that the project manager worries about small amounts and wants the precision. The semilogarithmic scale lacks this precision at the end of the project.

On the other hand, the semilogarithmic scale easily accommodates parallel percentage warning lines and danger lines. Since a semilogarithmic scale is a ratio scale, a line drawn parallel to it is a fixed percentage different from it. The project manager can thus easily draw warning lines, say ±5% of the budget line, and danger lines, say ±10% of the budget line. The actual percentages used depend, of course, on the amount of the budgetary contingency allowance. The warning percentage should be small compared to the total contingency allowance. The danger percentage should be, say, a third or half of the total contingency allowance.

A project manager (or a clerk) can plot the accumulated expenditures on the same semilogarithmic graph as the budget, warning, and danger lines. When the accumulated expenditures fall off the budget line but within the warning lines, the project manager need not be too concerned. The contingency allowance is sufficient to cover such discrepancies. However, when the accumulated expenditures fall outside the warning lines, the project manager should find out what is going on. The discrepancy between budgeted and actual expenditures is getting large enough to suggest that something is awry. Moreover, it is growing in absolute terms if it just stays at the same percentage above or below the budget line.

E. Overspending

When accumulated expenditures exceed the upper danger line, the project manager may have a serious problem to resolve: A significant portion of the contingency allowance may be depleted before the project is finished. This requires the project manager to reprogram the remainder of the project in order to conclude it without overrunning the total budget.

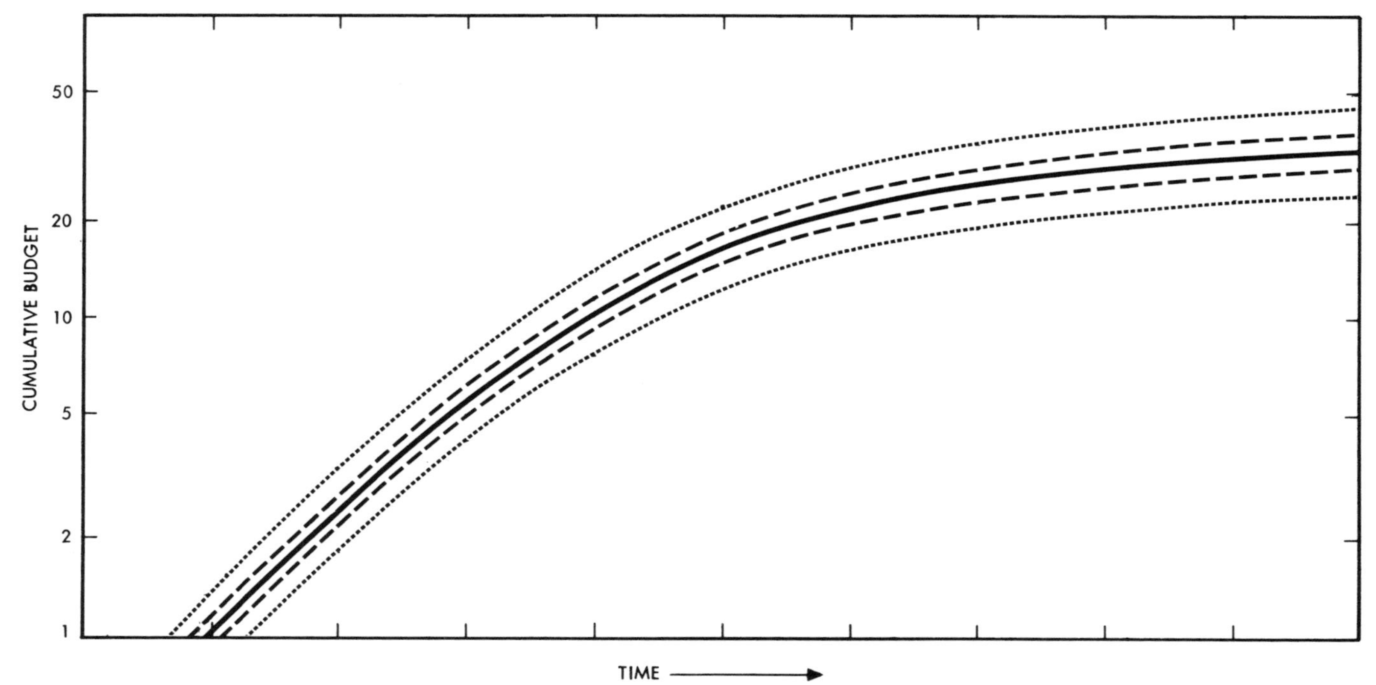

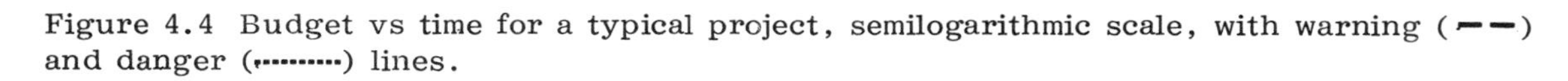

Figure 4.4 Budget vs time for a typical project, semilogarithmic scale, with warning (— —) and danger (⋯⋯) lines.

F. Underspending

So far, one might have assumed that an excess of accumulated expenditures over budget is more serious than a deficiency. However, a deficiency, especially in the early or middle phases of a project, can indicate significant trouble: If the budget is not being spent, then the work is probably not being done. Not only does this endanger completion of the project on time, but it also threatens the budget because a compressed work schedule almost always entails inefficiencies and increased costs.

G. Budget Needed to Complete the Work

Another expenditure/budget comparison is both the most useful and the most difficult to make, viz., a comparison of the budget yet unspent with the work yet to be done. This comparison is truly the most useful because it indicates whether the work remaining can be done within the available resources. The other comparisons are in fact retrospective and require the project manager to infer what future conditions will be by examining the past. A comparison of the work to be done and the budget yet unspent is more direct in terms of helping the project manager gauge and influence the future. Thus, this comparison is more helpful than the others. At the same time, it is difficult to use because it repeatedly requires the project manager to re-estimate what is needed to complete the project as the project progresses.

Each of the expenditure/budget comparisons has its advantages and disadvantages, and wise project managers will use each of them appropriately. Accumulated expenditures will likely be the easiest to apply overall and will probably serve as the basic approach. Periodic expenditures are useful for tracking selected aspects that have a tendency to go awry. And comparisons of work to be done versus budget yet unspent will help project managers restructure their plans if necessary.

5

Coordinating and Directing Techniques

Effective coordination and direction of the project team depend essentially upon successful motivation and good communication. Successful project managers understand what makes their team members tick and how to communicate with them productively. Accordingly, our approach in this chapter is to present principles of motivation and effective communication and show how they can be applied in project activities such as giving and receiving criticism, conducting meetings and conferences, holding conversations, and preparing written communications. Further information about motivation and communication can be found in *What Every Engineer Should Know About Human Resources Management.* *

We also introduce the subject of leadership styles. There is more than one way to lead and each may have its own special advantages, especially in terms of what motivates different team members under different circumstances. Project managers can enhance their chances of success by leading in ways that are particularly suitable for the situation at hand.

---

*Martin, D. D., and Shell, R. L., *What Every Engineer Should Know About Human Resources Management*, New York: Marcel Dekker, Inc. 1980.

I. MOTIVATION

The most potent force that a project manager can have to elicit cooperation from team members is to genuinely care for them. This is not "manipulating." Rather it is behavior that serves mutual best interests. When the project manager takes the initiative in caring for team members, he or she can expect them to reciprocate by caring about the project and its manager and cooperating within their ability to do so. This is also true in the project manager's relationships with superiors, peers, support staff, and customers.

Caring can be exhibited in many ways. Common everyday courtesies indicate caring and should be extended no matter what the other's station. Caring is also shown in conversation when the listener pays attention to the speaker and tries seriously to understand the speaker's point of view and feelings. When in a group, caring is shown by those who solicit the views of others who have not been able to insert themselves into the discussion.

Another aspect of human interaction is the practice of paying compliments. Compliments are positive comments or strokes and enhance the other person. As a result, the person who is complimented feels good about himself or herself and about the person who paid the compliment. He or she is thus inclined to cooperate with that person if possible. Of course, compliments must be sincere. If they are not, their insincerity will be detected and cooperation may in fact be withheld.

There are two kinds of positive strokes, conditional strokes and unconditional strokes. A conditional stroke is a compliment that can be paid because of something the recipient did. It is called a "doing" stroke. An example is "You did a nice job on that report."

An unconditional stroke, in contrast, is a compliment that can be paid just for being, and no strings are attached. It is called a "being" stroke. Examples are "It's good to see you" and "I'm glad to have you on the project."

Of the two types of positive strokes, unconditional strokes are the more potent. That is, they affect one's sense of self-worth more deeply and pervasively.

While discussing stroking, it is appropriate to caution against certain counterproductive comments. Negative strokes are putdowns and diminish the other person. An example might be "We'd like to pay you what you're worth, but the minimum wage law won't let us." Negative strokes are resented and build animosity. They should never be used. Negative strokes are not to be confused with criticism, which is discussed in Section III.

Closely related to negative strokes are blurred strokes. A blurred stroke leaves the recipient wondering what was meant. Was the stroke a compliment or a putdown? An example is "Your work isn't nearly as bad as they said it would be." There is always a negative quality to a blurred stroke, and it should not be used.

Similar to blurred strokes are crooked strokes, which also have a negative component. They appear on the surface to be compliments but they are actually very sarcastic putdowns. An example is "You look nice today; your socks match." Crooked strokes should be avoided. Even if one receives a crooked stroke, one should not return it.

Finally, the project manager should be aware of the consequences of ignoring an individual, i.e., not paying that person any strokes. People who receive no strokes are likely to foul things up just when it hurts most. They do this to get attention. Their behavior is similar to that of a child who does not normally receive some attention. The child then manages to get attention by misbehaving.

The unfortunate consequences that can result from non-stroking mean that the project manager cannot afford to focus attention on only poor performers. If satisfactory and outstanding performers are ignored, one or another of them will foul up just to receive a share of the manager's attention.

If stroking patterns among a project manager's team members are unhealthy, they can and should be changed. The best way for a manager to do this is to personally practice caring behavior, including positive stroking. Not only will this caring behavior enhance the team members' feelings about themselves, but it will also set an example for them to emulate in dealing with each other. In addition, the project manager can help improve stroking patterns by accepting a positive stroke graciously whenever one is paid and by not stroking another's bad behavior.

It was mentioned above that compliments must be sincere and paid genuinely if they are to be accepted as evidence of caring. Such sincerity grows out of an overall fabric of straightforward communication. One should not expect a compliment to be taken seriously if everything else that one says is fuzzy or misleading.

Straightforward communication, sometimes known as straight talk, occurs when the speaker (or writer) says what he or she means. For example, straight talkers say "I" when they mean "I" and do not hide behind vague "we's" or "they's." They are specific when they can be. If they need something by Tuesday noon, they do not say "anytime next week" (and then perhaps become upset if the item is not ready by Tuesday noon). If they mean "no," they say "no" at the time. They do not just merely discourage the other person, hoping that he or she will get the point without being told "no." And straight talkers ask for what they want, rather than hint about it while hoping that the listener will deduce what they want.

II. COMMUNICATION

Communication can occur in three ways:

1. through words *per se*, as in a memo or letter
2. through vocals such as tone, emphasis, hesitations, etc., as in a telephone or face-to-face conversation
3. through non-vocals or body language such as facial expressions, gestures, position, etc., as only in a face-to-face conversation

Generally, a face-to-face conversation is richer in the amount of information transferred than a telephone conversation, which is richer than a written communication. This is particularly true when feelings are involved, which is often the case in project work. For example, a lack of belief, commitment or a lack thereof, a defensive attitude, hopefulness, enthusiasm, and a host of other feelings that affect project performance can be detected when vocals and non-vocals are observed; they might not necessarily be detected from words alone. It behooves project managers, therefore, to communicate face-to-face and to urge their staff members to do likewise whenever serious issues are at stake.

It is also important for people who are working together for the first time to meet face-to-face and become acquainted. Then when they converse over the telephone they will have an idea of how the other's vocals might correspond to their unseen non-vocals. This will enable them to read the other person a little more accurately.

To take good advantage of the available vocals and non-vocals, one must be an attentive listener, seeing as well as hearing what the other is transmitting. Unfortunately, some people are regularly poor listeners:

1. They are so busy framing their replies that their thinking interferes with their seeing and hearing. If thoughts toward

a reply do occur while they are listening, they should make a note or two for later reference and resume listening.

2. They feel dull and bored, unable to attend to each word. Bored listeners can sometimes help break the dull pace by asking appropriate questions.
3. They lack self-confidence and worry about the impression they are making, or are about to make, so that they do not pay attention to the speaker. This is most likely to happen in formal settings, where each person knows when he or she is due to speak. The best remedy for this is to practice in advance so that there is little doubt about one's ability to perform satisfactorily.
4. They dismiss the communicator as unimportant because he or she seems unable to hurt or help them. Cavalier dismissal of communicators can be dangerous. Perhaps these communicators will fade into the background and not volunteer information crucial to the project's success.

Sometimes one detects that others are listening poorly. In this event, it is appropriate to ask the poor listener a question that requires a response and wait for it. This can be done without putting anyone down and will usually bring all the listeners back into the conversation.

Good listeners also have some clear characteristics:

1. They can repeat back what has just been said. Beware, though, that false assurances may be gained if the non-vocals are missed.
2. They see the body language that accompanies the verbal message.
3. They detect minor discrepancies, small gaps or omissions, and minute ambiguities. Sometimes discrepancies, gaps, omissions, and ambiguities occur between different conversations. Questions can be asked to clarify the situation when they are noticed, unless a question would be distracting. Then it is better to probe later. But such little aberrations may tip off important information. What has happened, for example, to make the speaker more optimistic? Or pessimistic?
4. They notice when words, phrases or expressions are used in an unusual or curious way, which may signify a critical group experience. A critical group experience may have been either very pleasant or very unpleasant. Until it is known which type it is, one should steer clear of referring to it lest one upset a member of the group.

5. They are not embarrassed to ask for a moment to think when they have been too busy listening to frame a reply. One should feel free to ask for time to think before responding. It is courteous and shows that the listener has indeed been listening.
6. They habitually find value in everyone who seeks to communicate with them. Such listeners will find that they are among the first to know what is going on, rather than among the last.

III. CRITICISM

From time to time, a project manager will need to criticize the work of a team member, a supporting element, or even the boss in order to improve some performance. The art of giving criticism should be mastered carefully, for upon it hangs much of the project manager's ability to bring a project to a successful conclusion.

There is only one legitimate purpose of criticism, and that is to change behavior or performance. The presence of any other objective, e.g., to blame or shame another, is both inappropriate in a caring relationship and counterproductive. People simply do not feel like changing to accommodate someone who has just belittled them.

Since change is the object of criticism, an assessment must be made of the ability of the recipient to change. Sometimes change is not possible. The reasons may range from the desired change lying beyond the other's talent or knowledge, to his or her being psychologically unable to change, to the desired change being forestalled by matters outside one's control. Whenever the recipient is unable to change, regardless of the reason, there is no point in giving criticism. In these cases, the criticism only frustrates the recipient and does not produce the desired result.

Deciding whether another person is capable of changing is not necessarily easy, nor is one's judgment necessarily going to be correct. This does not mean that the would-be critic should play it safe and refrain from criticizing. Rather it means that one should try earnestly to assess the situation first and proceed gradually if one believes but is not sure that the recipient is able to change. By proceeding gradually, one can get some feedback before coming down too hard on the recipient if in fact the preliminary assessment is wrong. A caring disposition plus a normal amount of sensitivity to the other person will almost always enable a would-be critic to begin at an appropriate pace, read the signals given by the other, and proceed if on the right track.

Sometimes the recipient of criticism should be asked what would help him or her make the desired changes. Perhaps the situation can be negotiated, so that the critic receives what the critic truly needs and the person being criticized can perform in a way that he or she finds acceptable.

Criticism should be specific and in terms of the desired behavior or performance. If necessary, an example of acceptable behavior or performance should be provided. Generalities should be avoided, for they tend to divert the recipient's attention away from the desired change. Since a generality in criticism is usually accusatory, the recipient may search for an exception to disprove the accusation and argue over the past rather than focus on changing for the future.

The project manager is sometimes the recipient of criticism from one who is not very skillful. If so, he or she can help make the exchange productive by:

1. viewing the criticism as an opportunity to learn what is important to the critic
2. drawing out the critic to obtain specific information on the changes desired
3. watching non-verbal cues to see if the real substance of the critic's message is the same as the words
4. not entering into needless arguments

IV. MEETINGS AND CONFERENCES

An important vehicle for coordinating and directing project activities is the meeting. Most project personnel are familiar with meetings, but in their experience good meetings may be the exception rather than the rule. Successful project managers will correct any tendencies toward unsatisfactory meetings and make their meetings productive.

Successful meetings depend upon three factors:

1. task management
2. process
3. physical arrangements

We discuss each of these factors below, as well as two factors that lie on the interface between task management and process, viz., agendas and minutes.

A. Task Management at Meetings

Task management refers to the business of the meeting, i.e., the accomplishment of a specific task or set of tasks which is the purpose of the meeting. Accordingly, task management begins when one decides on the goals of the meeting.

Potential meeting goals are:

1. to define a problem
2. to create alternative solutions to a problem
3. to weigh courses of action
4. to decide among alternatives
5. to plan future acitvities
6. to exchange information
7. to review and clarify, i.e., to make sure everyone understands
8. to evaluate progress
9. to raise issues

Sometimes it is appropriate to pursue just one of these goals at a single meeting. At other times it may be appropriate to pursue two or more, such as goals 1 through 4 or goals 6 and 8.

Project managers and other meeting attendees should realize that some participants may have hidden meeting objectives which may interfere with accomplishing the stated goal. Typical hidden objectives are:

1. To block, delay, or confuse action. Some people will use a meeting to muddy the waters because they do not like the course being followed.
2. To fill up time, avoid other work, or get a free lunch.
3. To fill a meeting quota. Some meetings are scheduled to occur periodically whether they are needed or not. However, this arrangement may be valid if it is important for the participants to touch base regularly.
4. To keep the group divided. When combatants are brought together, they are likely to harden their positions so that compromise cannot be achieved.

5. To diffuse decision responsibility so that it will be hard to blame anyone in particular for a bad decision.
6. To meet the social needs of the group. This is a legitimate purpose but it should not be confused with efforts to accomplish anything else.
7. To use the meeting as an arena to impress the customer or boss.

Any of the normally hidden objectives can be a legitimate meeting goal at one time or another. However, difficulties arise when different individuals have conflicting goals or when an individual's objective is at odds or cross purposes with the primary purpose of the meeting. In these cases, hidden objectives should be "smoked out" and confronted. The meeting goals that will be pursued will have to be decided and agreed upon before any useful work can be done.

Related to the goal of the meeting is the scope of the meeting. "Scope" refers to the extent to which the subjects will be explored or developed. It makes a difference, for example, whether a meeting to develop budgets will be concerned with budgets to the nearest hour of effort or to the nearest $10,000. Likewise, it makes a difference whether a meeting to consider oil drilling is to be concerned with all of North America or just the off-shore region of California.

Once the purpose and scope of the meeting are determined, it is easy to decide who should be invited. Conversely, if the purpose is not known, it is impossible to decide whom to invite.

The list of invitees should include those who can contribute toward the meeting's goals, and only those people. This does not mean that those who have little to say on the subject but need to hear should be excluded; indeed, they should be included. But it does mean that the meeting should not include idle observers who do not need to hear the proceedings. Their presence merely dilutes the attention of the real participants and it may also inhibit their candor.

Having determined the goal, scope, and participants for the meeting, it is appropriate for the leader and the participants to determine what homework they must do in order to perform their individual roles at the meeting. The invitees should understand the goal and scope well enough that they can easily determine what homework they must do.

Knowing the meeting's purpose, scope, participants, and the participant's needs for preparation, the chair can then schedule the meeting. The meeting must be held soon enough for its result to be timely and late enough for the participants to be able to prepare. It must also be scheduled so that the group has enough time to do the work that has to be done.

If at all possible, the length of the meeting should be set in advance and the participants notified. This allows them to schedule their own time better. If it is useful to keep the meeting from running on and on, then it can be scheduled shortly before lunch or quitting time. Most people will conclude an otherwise interminable discussion in order to eat or go home.

Finally, the location of the meeting must be decided. Both the purpose of the meeting and convenience to the participants are factors. Sometimes the purpose dictates or influences the choice of meeting location. Perhaps it would be useful to meet at the job site, perhaps in the vicinity of files or other resources that might be needed. Perhaps there is a psychological advantage to be gained (or lost) by one of the parties if the meeting is held in a particular location. Perhaps the meeting should be held at a neutral location. Perhaps the meeting should be held away from possible interruptions and distractions. Whatever location is selected, it should be chosen consciously with due regard to the impact that it can have on the success of the meeting.

B. Meeting Process

Meeting process refers to the way the participants behave and interact at the meeting. A meeting's objectives are not met simply because the attendees assemble in a room. Rather, certain activities must occur, such as initiating the discussion, obtaining information, and achieving agreement, if there is to be any progress toward the goal.

Key process activities are:

1. *Initiating* The chair is responsible for starting the meeting and initiating discussion, but all of the participants can raise issues or introduce new topics when appropriate, and they should raise important items that are being overlooked.
2. *Information and opinion providing* All meeting participants have a responsibility to share relevant information and opinions, whether or not they have been directly asked.

After all, they were invited to the meeting for the contribution they can make toward its goals.

3. *Information and opinion seeking* In some meetings, not all the participants will volunteer what they know or think or feel. Those who are quiet should be drawn out. Some may even have to be coaxed.
4. *Encouraging* Venturing an opinion, offering information, or just asking a question represents a psychological risk for some meeting participants. These people need to be encouraged, and everyone at the meeting should present an encouraging rather than a discouraging attitude.
5. *Reality checking* Assumptions, opinions, and speculations offered in the meeting need to be checked against facts. Whoever has factual knowledge about the subject being discussed has a responsibility to verify or challenge assumptions, opinions, and speculations.
6. *Analyzing* From time to time it will be appropriate to analyze what a situation really is, what some information really means, what is known and what is unknown, etc. Participants can contribute greatly to the progress of their meetings if they will do appropriate analysis and share the results with their fellow attendees.
7. *Clarifying* It is occasionally necessary to make sure that everyone understands—and has the same understanding. When there are expressions of bewilderment or confusion, the participants should clarify the issues before proceeding.
8. *Consensus testing* As a subject is developed in a meeting, it is important to determine if the participants agree or not on its various aspects. Formal votes are usually not required. If one cannot tell from the participants' comments or facial expressions, a simple question such as "Do we all agree on this?" may be all that is needed.
9. *Harmonizing* Points of agreement should be searched for and disagreements should be harmonized or resolved wherever possible. Whoever sees a way to reduce a disagreement should help do so by identifying points of similarity, alternative interpretations, possible compromises, and so forth.
10. *Summarizing* It is important to summarize a discussion after it has been underway for some time, or just before moving to a new topic even if the discussion was not particularly long. This provides the entire group a

compact statement of the key issues and identifies any loose ends that might exist. The group can then move forward with a sense of having reached a milestone. Also, the participants will have a common statement of what remains to be done, if anything.

11. *Recording* Meeting discussions and gains made should not be lost or subject to faulty recall in the future. Therefore, the chair should see that at least one person takes notes of the meeting. Alternatively, major items can be written on a flip chart or chalkboard for all to see and then later transcribed. Meeting records are discussed further in Section C.2.
12. *Goal tending* Goal tending refers to keeping the discussion on the topic and curtailing useless or untimely excursions. All meeting participants have a duty to tend the goal and to retrieve and refocus the discussion if it has strayed afield.
13. *Time keeping* Time keeping means avoiding unnecessarily protracted discussions and moving forward from one agenda item to the next as quickly as practical. As in the case of goal tending, all meeting participants have a duty not to waste meeting time and to help others avoid wasting meeting time.
14. *Gate keeping* Gate keeping means assuring that one or a few individuals do not dominate the discussion and thereby exclude others from participating. Again, participants must practice self-discipline so that they do not take up more than their fair share of the meeting time. They also have a duty to intercede if another attendee does not practice self-restraint.
15. *Listening and observing* If a meeting is not going well, whoever notices that something is amiss should listen to and observe the process in order to identify what is missing. He or she should then supply the missing element if possible or encourage others to provide it.

Some behaviors are productive at meetings and others are counter-productive. Productive behaviors include the following:

1. *Participating* People are invited to a meeting in order to participate in the discussion. If they are there just to listen, they might as well not come, generally speaking. They can read the minutes instead.

2. *Sharing strong feelings* When people say quietly to themselves or friends after the meeting, "I wish I had said . . .," they probably should have said it in the meeting. If they had, they might have found that they had company. Perhaps their comments would have changed the course of the discussion. Or they might discover that they are alone in their feelings and then wish to revise their positions.
3. *Using good timing* One should make reminder notes of ideas that properly belong later. Similarly, one should not wait until the end of the discussion to say something that should have been said earlier.
4. *Making sequence responses* A disjointed discussion is difficult to follow. To minimize this problem, participants should provide transitions between ideas already discussed and what they are about to say. In other words, participants should make apparent or obvious the sequential relations that exist between what they will say and what has gone before.
5. *Building on ideas* Other things being equal, it is more economical of meeting time if ideas already discussed are expanded, embellished, enhanced, and made better rather than displaced by new ideas. Improving and supporting another's idea helps consolidate the participants and move them toward agreement, while introducing new alternatives needlessly may divide the group and make agreement more difficult to reach.
6. *Paraphrasing to show understanding* Just as in conversations, it is helpful to check one's understanding by paraphrasing what one believes to be the case.
7. *Giving non-verbal as well as verbal cues* Facial expressions and other non-verbal cues help participants know if they are communicating clearly, have support for their ideas, are boring their listeners, and so forth without having to poll the assembly. This enables speakers to adjust their comments to the listeners' needs and expedite the flow of the meeting.
8. *Having courage and taking appropriate risks* It takes courage at times to express views that may not be popular, to question another's assertion, to ask for time to analyze the situation to bring another back to the subject, etc. Yet such courageous acts may be keys to advancing the progress of the meeting. Meeting participants should have courage and take appropriate risks in expressing themselves and in dealing with other participants.

Counter-productive behaviors in meetings include these acts:

1. *ego-tripping*, which is at the expense of everyone else's time
2. *inappropriate entertaining*, i.e., entertaining which disrupts the meeting and does not contribute toward either the task or process of the meeting
3. *subgrouping*, where a little pocket of people holds its own meeting, both withdrawing from the main meeting and disrupting it at the same time
4. *withdrawing* mentally from the meeting
5. *blocking*, i.e., obstructing the flow of the meeting, being argumentative for its own sake, and so forth
6. *dominating*, which not only takes up other people's time but prevents them from making their contributions
7. *sidetracking* or diverting the meeting from its purposes

As noted throughout this section, meeting process involves all the attendees. Each one is responsible for the success of the meeting. Each one should perceive what is going on in the group, identify what is missing in either process or task, and supply it if possible or see that it is supplied.

The meeting chair has additional duties which are not shared with the rest of the participants: The chair sets the stage for the discussion and introduces the meeting. The chair also introduces the participants to each other when they have not met before.

At the end of the meeting, the chair should thank the participants. The chair needs also to be sure that the participants receive a list of action assignments. Each entry in the list should include the name or description of the assignment, the assignee, the date due, the form of any deliverables, and the recipients of deliverables. If meeting notes or minutes are to be circulated (see Section IV.C), the chair also should ensure that this occurs in a timely way. The assignment list may then be incorporated in the minutes or appended to them.

C. Agendas and Minutes

Agendas and minutes lie at the interface between task management and meeting process. They are connected at once to both the purpose of the meeting and the conduct of the meeting.

1. Agendas

In simple terms, an agenda is a list of things to be done. Often printed agendas, especially those distributed publicly, are merely lists of the topics to be considered at the meeting. A meaningful agenda, however, shows not only the topic or name of the issue but also its goal and scope, the key participants, and some notion of the amount of time to be devoted to it. Meetings often flounder without an agenda, and it is generally wasteful to proceed until an agenda is agreed upon. The chair is responsible for drafting an agenda.

It is useful for the participants to review the chair's proposed agenda and modify or revise it as appropriate as their first item of business after introductions. This approach has three benefits. First, it may result in a sequence that makes better use of everyone's (or at least most peoples') time than the original agenda. Second, it helps the participants understand each other's priorities and interests. And third, it gives the participants a chance to trade small concessions and begin to build good will toward each other (see Chapter 7).

2. Minutes

Minutes, i.e., a record of the important aspects of a meeting, should be made of every meeting. Their form and formality may vary, depending upon the circumstances.

All minutes should include the following information:

1. the time and place of the meeting
2. the attendees, including part-time attendees and guests
3. a copy of the agenda (as an attachment)
4. a brief preamble on the purpose of the meeting
5. notes on issues discussed, including concerns, assumptions, decisions made, and open issues
6. action assignments (see Section IV.B)
7. when and where the next meeting will be, if known

If formal minutes are called for, they should be organized, edited for clarity and accuracy, typed, and distributed within a few days of the meeting. The attendees should receive the minutes while their memories are still fresh. This will enable them to review the minutes meaningfully and identify any errors of omission or commission. If minutes are received after memories have dimmed or become selective, needless and perhaps unresolvable arguments can develop about what really transpired at the meeting.

Advantage accrues to the one who prepares the minutes. There is usually more than one way to express what happened at the meeting, and the individual who prepares the minutes can choose the words. In time, i.e., after the minutes are formally approved if that practice is followed or after no one complains if that is the practice, then the minute taker's record will be the only official record. Thus, those who refer to the official record later will have his or her version of the proceedings.

D. Physical Arrangements for Meetings

The success of the meeting may depend upon its physical arrangements. While satisfactory arrangements cannot guarantee success, unsatisfactory arrangements can almost certainly assure a more tiring and less productive meeting.

The following conditions are essential for good meetings:

1. comfortable air quality and temperature
2. ability of the participants to see and hear each other without straining
3. freedom of the participants to stretch, get up, move about, etc. without disturbing others
4. soft but adequate lighting and no glare from windows, walls, chalk boards, etc.
5. proper furniture

V. FACE-TO-FACE AND TELEPHONE CONVERSATIONS

Not all project interactions require or indeed are well-served by meetings. Face-to-face and telephone conversations on one hand and written communications on the other are alternative ways to conduct project business. Conversations are discussed in this section and written communications are discussed in the next. In both cases, attention is focused on the purposes that each can serve and on ways to make them effective.

A. Purposes of Conversations

Conversations are two-way exchanges and are best used where it is important to know how well one is communicating while proceding or where one of the parties needs some response before proceding. Thus, conversations can be effective used for the following purposes:

1. to give information and instructions
2. to obtain information
3. to coach
4. to seek cooperation
5. to keep in touch

Information and instructions can be given by written communication, as discussed below, but there are times when a conversation is better. Whenever the information or instruction is subject to partial or alternative interpretation, depending upon the background and understanding of the recipient, it is helpful to communicate orally. This way, the sender can check periodically to verify that the receiver has grasped the information or instructions as they were intended. The sender can observe vocals and non-vocals as well as ask the receiver to paraphrase to ascertain understanding. Once understanding has been achieved, it can be confirmed in writing for the record if necessary.

Likewise, information can sometimes be obtained most readily through a conversation. This is particularly true where the request is subject to being misinterpreted or where the scope needs to be developed by the requester and the responder together before the request can be addressed. Again, the request can be confirmed in writing once it has been adequately developed through conversation.

Coaching consists of giving instructions one at a time where each instruction is tailored to the recipient in light of the recipient's response to the previous instructions. Coaching cannot be effectively done in written form.

Cooperation is also better developed by conversation rather than written communication. When seeking cooperation, one needs to be aware of the other person's responses in order to know best how to unfold the request. Also, the person whose cooperation is being sought generally finds it more difficult to ignore a face-to-face or telephone request than to a written request. Moreover, a conversational request affords an opportunity to negotiate the exact form or content of the cooperation. For all these reasons, a conversational request is more likely to elicit cooperation than a written request.

Keeping in touch is an important part of project management, whether the person to be kept in touch with is the customer, task leaders, support function personnel, subcontractors and vendors, or the boss. Project managers typically can use as much insight as they can get into other peoples' intentions and concerns. This insight will help them prepare for possible changes in goals or plans. If they wait until the changes have such a high status that they are written down, then they may find it difficult to accommodate them. On the other hand, if they get early indications of possible changes, then they can prepare for them with less difficulty. Keeping in touch establishes and maintains an open communication channel for the parties to exchange ideas without having to organize their thoughts into self-consistent expositions and without committing the thoughts to writing. People will often share notions with others if they can do so effortlessly and without worrying whether they will be held to them. Conversation promotes such sharing, while writing inhibits it.

B. Preparing for a Conversation

One should prepare for all but impromptu conversations the same way one prepares for a group meeting. One needs to know the goals and the scope of the discussion and needs to do whatever homework may be necessary. Also, one needs to consider how to begin the discussion so as to encourage the other person to continue the conversation rather than end it.

C. Conducting the Conversation

A productive conversation consists of several distinct steps. Often they seem to occur effortlessly. However, no one should be deluded into thinking that they will occur automatically. Some attention must be paid to conducting the conversation by at least one of the parties to the conversation. Project managers will enhance the conversations that they initiate if they take responsibility for the following steps:

1. *Establishing the purpose of the conversation* The project manager should let the other person know the ostensible reason for the conversation. One should not have to wonder about or guess the purpose. It is true, however, that the project manager may wish to reveal only part of the purpose at the outset. The project manager may want to wait until he or she has some insight into the other person's viewpoint or frame of mind before revealing the rest of the purpose. If an invitation is extended or an appointment is made to hold the conversation in advance of the conversation itself, it is appropriate to convey the purpose and scope of the conversation then. This practice will enable the other person to prepare for the conversation if necessary.
2. *Obtaining the other party's interest* Directly related to establishing the purpose of the conversation is obtaining the other party's interest. The other party must feel that it is more worthwhile to hold the conversation than to do anything else at this time. Unless the purpose itself makes the significance of the conversation patently clear, it is probably necessary to explain why the topic is important enough to warrant the other party's time and attention.
3. *Exchanging information* Both parties are more likely to consider the conversation productive if they exchange information in small increments. First, this practice allows each to determine how much the other knows about the subject so that his own comments neither assume too much nor are needless. Second, it allows each to participate actively in the conversation so that each one's interest is maintained. And third, it allows each to tailor what each says to the other's mood and thereby avoid both creating a bombshell and missing the target.
4. *Assuring understanding* Each party in a conversation needs to be understood and to know that he or she is understood. If clear and complete understanding is not evident from the other's remarks, one should ask the other to confirm understanding by paraphrasing what has been said. Likewise, one can volunteer to paraphrase to show one's own understanding. If a difference of understanding appears, the issue should be explored or developed further until the two parties agree. Whatever disagreements remain should be

about preferences, opinions, or values and not about facts or what transpired in the conversation.

5. *Obtaining commitments and establishing follow-through assignments* When one party to the conversation wants the other to do something, he or she must obtain the other's commitment to do it. Many times either or both parties will be ready to agree to do something but the commitment or agreement itself is not vocalized. It is important, however, for these agreements to be made explicit. Both parties should state unequivocally what they are going to do and by when they will do it. It is often a good idea for one of the parties to confirm their agreements in writing. And who shall write the confirming memo shall be one of the items agreed upon.
6. *Recording salient details* Wise project managers will always prepare their own records of salient conversational details. They may or may not provide a copy to the other party. These records should include not only information on what transpired but also notes on lingering concerns, possible threats and opportunities, and other "intelligence" that might later be useful.

VI. WRITTEN COMMUNICATIONS

Written communications are an alternative to meetings and face-to-face and telephone communications. Like the other forms of communication, they are better in some applications than others. As in the previous section, we discuss the purposes which written communications serve well and ways to use them effectively.

A. Purposes of Written Communications

Written communications are most useful when precise expression is important, when a dialogue is not needed, or when the recipients are so dispersed that oral communication is quite impractical. They should not be relied upon to coach, to elicit cooperation, or to keep in touch. Thus, written communications can be used effectively for the following purposes:

1. to give or request information and instructions that do not involve interactions between the parties

2. to issue reminders
3. to confirm, record, or document earlier conversations

B. Style

Style is an important ingredient in effective writing. The ultimate measure of style is whether or not it helps the reader understand the writer's message. Styles that detract from understanding are bad and styles that help are good. Thus, the preferred style depends partly upon the readers and their ability to understand from the style used. Nevertheless, a general axiom or two will serve all writers well.

Perhaps the most important advice is to put first things first. The subject and its importance to the readers should be stated at the outset. Project personnel are busy people and they will not wade through a memorandum or letter to determine if they should have read it. Unlike a conversation, the writer has only one chance to capture the reader's interest. If one does not seize the reader's attention immediately, one might as well not send the communication at all.

Also, it is important that the more important points of the communication be presented before the less important points. If the reader should get only part way through the communication, what he or she has already read should be more important than what he or she will miss by not reading the rest.* Moreover, truly significant elements should be highlighted and not buried in continuous prose.

It is also important that the writing be clear and simple. Long sentences, perhaps composed of subordinate clause upon subordinate clause, should be avoided. Try to communicate to rather than impress the reader. Where possible, use a phrase instead of a clause, and use a word instead of a phrase. While keeping to simple sentences, avoid monotony. Read your writing aloud to see if it flows without being a torrent.

---

*Suggesting that the most important material be placed first assumes that all potential readers have similar priorities. If they do not, then they may need different versions or at least a guide in the introduction to tell them where they will find what most interests them.

C. Content of Written Communications

Written project communications should contain what the readers need to know, and only that. A memorandum or letter should not be cluttered with extraneous material. (Readers should also watch for what might have been said but was omitted. There may be significance in the omission.)

A written communication should also tell its readers what the information means and why it is important. This should not be left to their imagination.

Finally, the communication should tell the readers what action, if any, they should take and when. It should also tell them if a response is expected.

D. Memo and Letter Organization

There are many possible ways to organize a lengthy, complex memorandum or letter. The following list of eleven ways can be used as a suggestion menu if one has trouble identifying a possible approach. Sometimes two or more ways can be combined. One way, say geographical (Number 2), can be the primary organizing framework and another, say chronological (Number 1), can be an organizing framework within each primary element.

1. chronological
2. geographical, e.g., site A, site B, etc.
3. from the known to the unknown
4. from the concrete to the abstract
5. from the simple to the complex
6. from the general to the specific
7. cause and effect
8. problem description, analysis, and solution
9. functional, e.g., by discipline
10. from the most important to the least important
11. questions and answers

Numbers (3) through (6), depending upon the circumstances, can be especially helpful when readers need to understand a complex situation. In the case of number (8), the solution can be put first if readers already have some idea of the problem.

VII. LEADERSHIP STYLES

Many theories and models of leadership apply to a greater or lesser degree to project management, and the reader is referred to *Handbook of Leadership** for a review of some major approaches. Three models, however, are particularly relevant to engineering project management because of the kinds of people and work involved, i.e., creative, responsible individuals who expect and are required to define and solve problems that are neither trivial nor familiar. These models are the Porter strength deployment model, the Myers-Briggs type model, and the Fiedler contingency model. We introduce them in this section to help project managers conceptualize and understand the human interactions that occur on projects. Through this insight, project managers can capitalize on favorable situations and mitigate the impacts of unfavorable situations. Moreover, they can use their insight to guide their own behavior in order to enhance overall project performance.

A. The Porter Strength Deployment Model

Elias H. Porter has developed a model of how different individuals behave when everything is going well and how they behave when they face opposition and conflict. If everything automatically went well in projects, we would not need to be concerned (nor need this book), for there would be no problems. Projects are, however, characteristically beset with problems, so it pays to understand how different people react in these conditions and how to make the best of the situation.

Porter identifies three primary motivations for human interactions; each person behaves in ways that represent some blend of these motivations. The differences among individuals are the different blends that we each have, like different fingerprints. Porter's primary motivations are:

1. *The altruistic-nurturing motivation* People who are altruistic and nurturing feel good when they are genuinely helpful to others. They generally are trusting, optimistic, loyal, idealistic, helpful, modest, devoted, caring, supportive, accepting, and adaptable.

---

*Ralph M. Stogdill, *Handbook of Leadership*, New York: The Free Press, 1974.

2. *The assertive-directing motivation* People who are assertive and directing feel good when they influence others. They generally are self-confident, enterprising, ambitious, persuasive, forceful, quick to act, imaginative, proud, bold, risk-taking, accepting of challenge, and eager to organize others.
3. *The analytic-autonomizing motivation* People who are analytic and autonomizing feel good when they are self-reliant, self-dependent, and self-sufficient. They generally are cautious, practical, economical, reserved, methodical, analytical, principled, orderly, fair, perserverant, conservative, and thorough.

In addition to these archetypical or primary motivations, Porter describes archetypical blends of the primary motivations taken two at a time in approximately equal parts.

4. *The assertive-nurturing motivation* People who are assertive and nurturing feel good when they are assertive in helping others. They generally are not autonomy-seeking.
5. *The judicious-competing motivation* People who are judicious and competing feel good when they are strategic, i.e., the power behind the throne. They generally are not nurturing.
6. *The cautious-supporting motivation* People who are cautious and supporting feel good when they are helpful to others but self-sufficient themselves. They generally are not assertive.

There is, of course, a seventh archetypical blend, which includes approximately equal parts of all three primary motivations:

7. *The hub motivation* People who have the hub motivation feel good when they do whatever the situation calls for. They generally are flexible and want to be included in whatever is going on.

The primary motivations and the archetypical blends can be represented by a triangular drawing, shown in Figure 5.1. The primary motivations are at the corners of the triangles; archetypical blends of two primary motivations are along the sides joining the two respective corners; and the archetypical blend of all three primary motivations lies in the interior of the triangle.

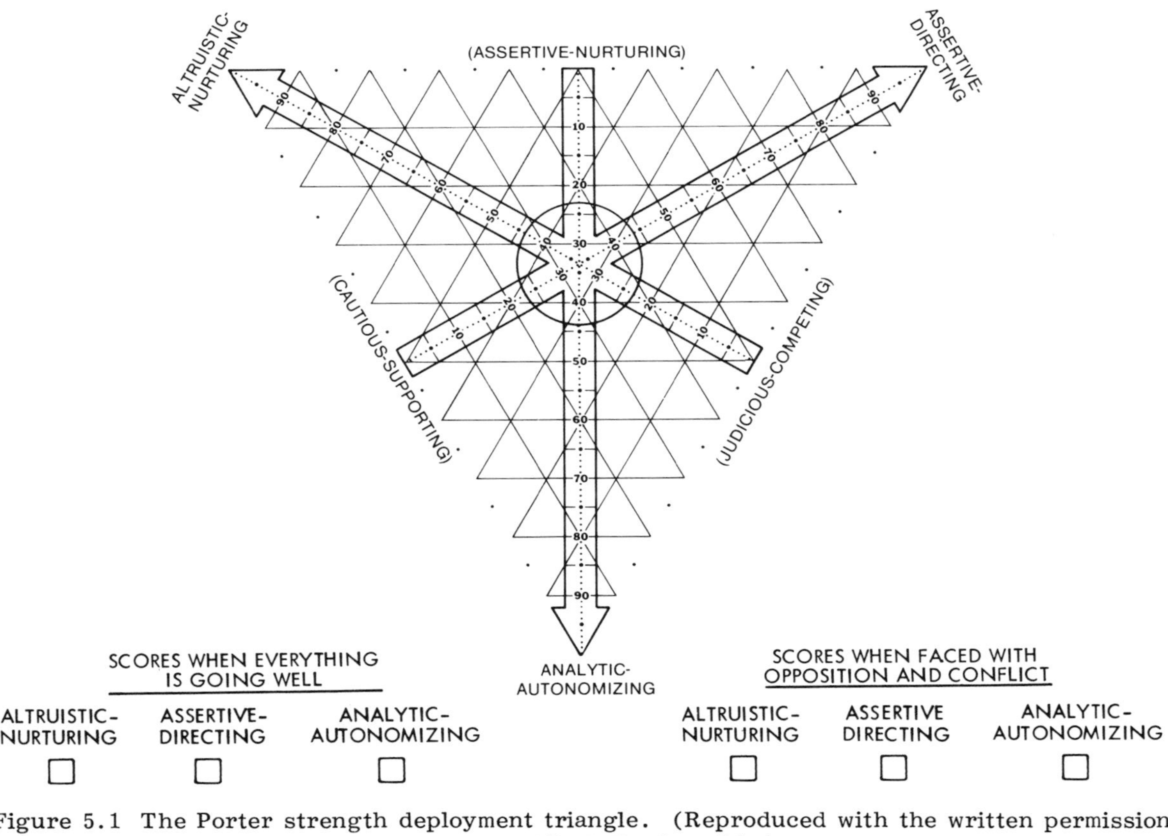

Figure 5.1 The Porter strength deployment triangle. (Reproduced with the written permission of Personal Strengths Publishing, Inc., Pacific Palisades, CA.)

The closer an individual's blend is to a given corner, the more dominant the corresponding primary motivation is overall. This relationship is shown by the numbers leading from the corner to the opposite side of the triangle. The number 100% coincides with the corner and the number 0% lies on the opposite side. It is no coincidence that the percentages for all three primary motivations total to 100 for every point in the triangle.

Porter has developed a paper and pencil test for individuals to use in determining their own motivational blends first when everything is going well and then when they are in conflict and opposition. Most people have different motivational blends in the two types of situations. The shift from one blend to the other can be represented by an arrow that points from the "going well" blend to the "conflict and opposition" blend, as in Figure 5.2.

People use their "going well" blends and their "conflict and opposition" blends differently. When everything is going well, people use their middle and lowest motivations and behaviors to support their pursuit of gratification from their highest motivation. That is, they use all their behaviors in a concerted way. Such a concerted approach facilitates their doing whatever is appropriate to advance the group's efforts toward their common objectives.

When faced with conflict and opposition, on the other hand, people use their motivations serially. Thus, they try to gratify themselves by using their preferred behaviors first. If their first approach is not successful, they try their next preferred behavior. If it too is not successful, they resort to their least preferred behavior, but they may do so reluctantly and possibly with vengeance (for being forced into such a low preference style).

The astute project manager will try to infer how the members of the project team are motivated and will behave and then use this insight in several productive ways. For example, the project manager can check whether the team has all of the primary motivations represented to some degree. Is there at least one person who will just naturally analyze a difficult situation, or one who is altruistic-nurturing and will naturally try to harmonize different viewpoints when necessary, or one who is sufficiently assertive and directing that the group will not flounder for lack of leadership? If a needed primary motivation is missing, then the project manager can either consciously try to fill the void(s) personally, structure meetings so that needed activities occur even if they are not quite natural for anyone there, or bring in resource people who will supply the needed ingredients.

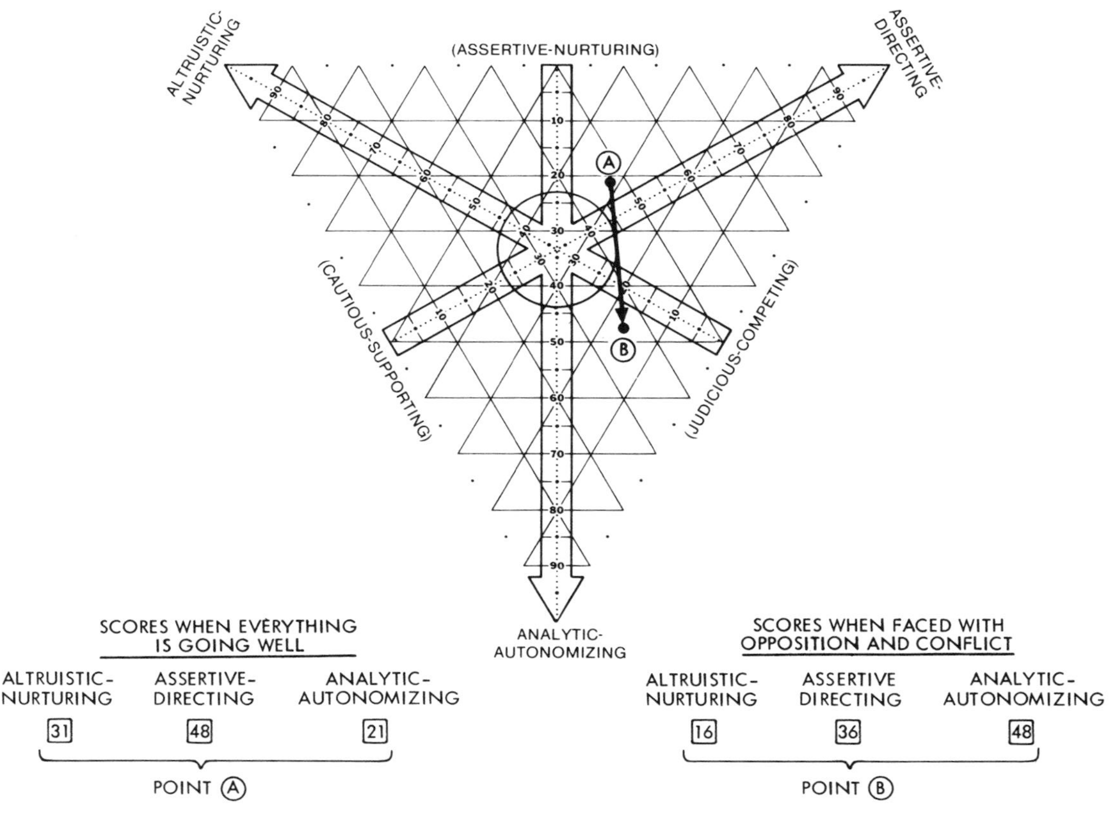

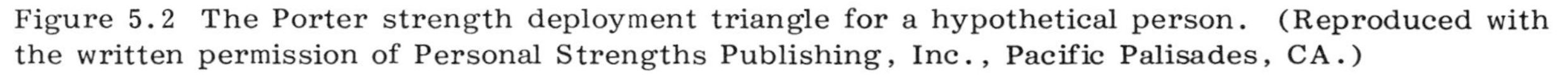

Figure 5.2 The Porter strength deployment triangle for a hypothetical person. (Reproduced with the written permission of Personal Strengths Publishing, Inc., Pacific Palisades, CA.)

Another possibility is for the project manager to use insight in selecting team members in the first place. Whenever the project manager has a choice, he or she can try to obtain all the needed behaviors, as well as the needed technical expertise, by selecting individuals according to their motivations.

A third use of motivational insight occurs when there are disagreements about what to do. The wise project manager will keep team members from having to act according to their least preferred motivation and thereby avert vengeful and unproductive behavior. Moreover, the project manager will try to avoid everyone's wasting time and effort by avoiding arguments that are basically rooted in individual motivations and approaches rather than in objectives. Thus, the manager will accept individual ways of doing things as long as they do not cause a problem elsewhere and will try to develop such forebearance throughout the project team.

Porter makes two observations about the use and perception of motivations. When a motivation is appropriately used, it is a strength. When it is overused, it is a weakness, which is to say that it is counterproductive. Thus, being analytical is a strength when analysis is called for, but being analytical is a weakness when it is time to harmonize or time to act. Similarly, being assertive-directing is a strength when direction is needed, but it is a weakness when the situation calls, say, for analysis. Being flexible is a strength when flexibility is needed; however, being flexible is a weakness when steadfastness is needed.

People who are motivated differently tend either to admire or to despise each other, rather than merely understand and accept them, depending upon whether mutual trust and respect exist. Table 5.2 gives some common ways that people of one set of motivations perceive people of other motivations.

B. The Myers-Briggs Type Model

Isabel Briggs Myers and Katherine Cook Briggs have considered human behavior in terms of how people perceive the world around them, how they make judgments or decisions, whether they prefer to perceive or to judge, and whether they share their preferred activity, i.e., perceiving or judging, with others. These differences can be important in understanding what goes on in project work. As in the case of insight gained from Porter's model, insight from the Myers-Briggs model will help the project

TABLE 5.1
Characteristics Strengths and Their Corresponding Weaknesses*

Altruistic-Nurturing Characteristics	Assertive-Directing Characteristics	Analytic-Autonomizing Characteristics	Hub Characteristics
Trusting Gullible	Self-confident Arrogant	Cautious Suspicious	Flexible Inconsistent
Optimistic Impractical	Enterprising Opportunistic	Practical Unimaginative	Open to change Wishy-washy
Loyal Slavish	Ambitious Ruthless	Economical Stingy	Socializer Can't be alone
Idealistic Wishful	Organizer Controller	Reserved Cold	Experimenter Aimless
Helpful Self-denying	Persuasive Pressuring	Methodical Rigid	Curious Nosy
Modest Self-effacing	Forceful Dictatorial	Analytic Nit-picking	Adaptable Spineless
Devoted Self-sacrificing	Quick-to-act Rash	Principled Unbending	Tolerant Uncaring
Caring Smothering	Imaginative Dreamer	Orderly Compulsive	Open to compromise No principles

(continued)

TABLE 5.1 (Continued)

Altruistic-Nurturing Characteristics	Assertive-Directing Characteristics	Analytic-Autonomizing Characteristics	Hub Characteristics
Supportive Submissive	Competitive Combative	Fair Unfeeling	Looks for options No clear focus
Accepting Passive	Risk-taker Gambler	Perservering Stubborn	Socially sensitive Deferential

*The first entry in each box is a characteristic strength. The second entry is the corresponding weakness, i.e., a strength overdone for the situation at hand.

Source: Reproduced from the *Leader's Guide to Relationship Awareness Training* TM by permission of Personal Strengths Publishing, Inc.

TABLE 5.2
Typical Ways that People Who are Motivated Differently Regard Each Other

1. An Altruistic-Nurturing Person regards:

   An Assertive-Directing Person as fearless and vigorous* or as ruthless and domineering**

   An Analytic-Autonomizing Person as idealistic and objective* or as cold and stubborn**

2. An Assertive-Directing Person regards:

   An Altruistic-Nurturing Person as supportive and warm* or as sentimental and weak**

   An Analytic-Autonomizing Person as informed and analytical* or as impractical and nitpicking**

3. An Analytic-Autonomizing Person regards:

   An Altruistic-Nurturing Person as intuitive and understanding* or as irrational, sentimental, and subjective**

   An Assertive-Directing Person as enterprising and bold in planning* or as unorganized and impulsive**

*Positive attributes, when mutual respect and acceptance exist.

**Perceived negative attributes which may or may not actually exist, when there is disrespect, dislike, or suspicion.

manager take advantage of opportunities and rectify shortcomings.

Myers and Briggs define the perception dimension in terms of whether people prefer to take in all the details or to see things whole, i.e., without focusing on the details, when they obtain information about the world around them. The first, i.e., the detail approach, is called sensing, and the second, i.e., seeing things whole, is called intuition. While most people do some of each, they nevertheless prefer to get information one way or the other when they have a choice. Behavioral traits that accompany perception preferences are listed in Table 5.3. Taking the entire lists as a whole, one can infer a person's preference, which can later be used in conjunction with the other dimensions to understand how the individual may behave in other settings.

TABLE 5.3
Perceiving Preferences

Intuitive Types	Sensing Types
Like solving new problems	Dislike new problems unless there are standard ways to to solve them
Dislike doing the same thing repeatedly	Like established ways of doing things
Enjoy learning new skills more than using them	Enjoy using skills already learned more than learning new ones
Work in bursts of energy powered by enthusiasm, with slack periods in between	Work rather steadily, with realistic ideas of how much time is required
Reach conclusions quickly	Usually reach conclusions step by step
Are impatient with routine details	Are patient with routine details
Are patient with complicated situations	Are impatient when the details get complicated
Follow their inspirations, good or bad	Are not often inspired and rarely trust their inspirations when they are
Frequently make errors of fact	Seldom make errors of fact
Dislike taking time for precision	Tend to be good at precise work

Source: Reproduced by special permission of the publisher, Consulting Psychologists Press, Inc., 577 College Avenue, Palo Alto, CA 94306, from *The Myers-Briggs Type Indicator Manual* by Isabel Briggs Myers © 1962.

Just as people prefer one way of perceiving, Myers and Briggs believe that people have a preference for the way they make judgments or decisions. The judging dimension is defined in terms of whether people prefer to decide on the basis of how things feel or on the basis of cold, logical analysis. These alternative approaches are called feeling and thinking, respectively. Again, most people decide sometimes on one basis and sometimes on the other, but given a choice, they have a preference for one approach or the other. Behavioral traits that accompany the judging preferences are shown in Table 5.4. As before, one can infer another person's preference by seeing which complex of traits that person seems to exhibit.

The Myers-Briggs model also includes a person's preference for perceiving or judging, as alternative activities. Thus, an individual not only has a preferred way of perceiving and a preferred way of judging but also has a preference for using one of these preferred activities over the other. A person who strongly prefers to perceive may find it difficult to make decisions, while one who strongly prefers to judge may find it difficult to observe the environment sufficiently to get the information needed in order to decide. Behavioral traits that accompany these preferences are listed in Table 5.5.

Finally, the Myers-Briggs model includes the dimension of introversion-extraversion. Extraverts reveal or share their preferred activity, i.e., perceiving or judging, with others. Introverts keep the activity that they prefer to themselves and share it only with their most intimate associates; they reveal, therefore, their second or less preferred activity. Behavior traits that accompany introversion and extraversion are listed in Table 5.6.

Table 5.3 suggests that intuitive and sensing types may behave oppositely. Indeed they often do, sometimes conflictingly and sometimes to good advantage. Conflicts can arise when one believes that the other should behave as oneself does. But if they could see how they complement each other, as in Table 5.7, then their differences could be very beneficial.

Likewise, Table 5.4 suggests that feeling and thinking types may behave oppositely. Again, their different behaviors can be conflictive if one expects the other to behave as oneself does. But, as before, if these opposite types could see how

TABLE 5.4
Judging Preferences

Feeling Types	Thinking Types
Tend to be very aware of other people and their feelings	Do not show emotion easily and are often uncomfortable dealing with people's feelings
Enjoy pleasing people, even in unimportant things	May hurt people's feelings without knowing it
Like harmony; may have their efficiency badly disturbed by office feuds	Like analysis and putting things into logical order; can get along without harmony
Often let decisions be influenced by their own or other people's personal likes and wishes	Tend to decide impersonally, sometimes paying little attention to people's wishes
Need occasional praise	Need to be treated fairly
Dislike telling people unpleasant things	Can reprimand or fire people when necessary
Are more people-oriented; respond more easily to people's values	Are more analytically-oriented; respond more easily to people's thoughts
Tend to be sympathetic	Tend to be firm-minded

Source: Reproduced by special permission of the publisher, Consulting Psychologists Press, Inc., 577 College Avenue, Palo Alto, CA 94306, from *The Myers-Briggs Type Indicator Manual* by Isabel Briggs Myers © 1962.

they complement each other, as in Table 5.8, their differences could be advantageous.

An astute project manager will minimize conflicts among opposite types of the project team by helping them recognize and appreciate the fact that together they can do better, more comprehensive and thorough work than either is likely to do alone.

If a project manager can determine what types are present among the project staff, he or she can identify behavioral traits that are likely to be missing in the group. The project manager

TABLE 5.5
Perceiving versus Judging Preferences

Perceiving Types	Judging Types
Adapt well to changing situations	Perform best when they can plan their efforts and follow their plans
Do not mind leaving things open for alteration	Like to get things settled and finished
May have trouble making decisions	May decide things too quickly
May start too many projects and have difficulty in finishing them	May dislike interrupting the projects they are on for more urgent ones
May postpone unpleasant jobs	May not notice new things that need to be done
Want to know all about a new job	Want only the essentials needed to begin their efforts
Tend to be curious and welcome new light on a thing, situation, or person	Tend to be satisfied upon reaching a decision about a thing, situation, or person

Source: Reproduced by special permission of the publisher, Consulting Psychologists Press, Inc., 577 College Avenue, Palo Alto, CA 94306, from *The Myers-Briggs Type Indicator Manual* by Isabel Briggs Myers © 1962.

can then try to compensate for missing ingredients by introducing some structure into the group's affairs. For example, if the group lacks an intuitive type, the project manager can ask one to review the project plan before it is launched in order to uncover previously unseen future possibilities. Or, if the group lacks a sensing type, the project manager can have the plan reviewed by one not on the project team who will check the soundness of the arrangements for coping with the immediate. If the group lacks a feeling type, the manager can consult with one about how the customer will feel. And if the group lacks a thinking type, the project manager can ask one to help the group analyze its situation.

TABLE 5.6
Concealing versus Sharing Preferred Activities

Concealing Types (Introverts)	Sharing Types (Extraverts)
Like quiet for concentration	Like variety and action
Tend to be careful with details; dislike sweeping statements	Tend to be faster; dislike complicated procedures
Have trouble remembering names and faces	Are often good at greeting people
Tend not to mind working on one project for a long time uninterruptedly	Are often impatient with long slow jobs
Are interested in the ideas behind their jobs	Are interested in the results of their jobs, in getting them done, and in how other people do them
Dislike telephone intrusions and interruptions	Often do not mind the interruption of answering the telephone
Like to think a lot before acting, sometimes without acting	Often act quickly, sometimes without thinking
Work contentedly alone	Like to have people around
Have some problems communicating	Usually communicate freely

Source: Reproduced by special permission of the publisher, Consulting Psychologists Press, Inc., 577 College Avenue, Palo Alto, CA 94306, from *The Myers-Briggs Type Indicator Manual* by Isabel Briggs Myers © 1962.

C. The Fiedler Contingency Model

The Fiedler contingency model is based upon the idea that group performance is contingent or depends upon the circumstances,

TABLE 5.7
Sensing and Intuitive Types Can Supplement Each Other

A Sensing Type Can Help An Intuitive Type By	An Intuitive Type Can Help A Sensing Type By
Bringing up pertinent facts	Bringing up new possibilities
Applying experience to problems	Supplying ingenuity on problems
Reading the fine print in a contract	Reading the signs of coming change
Noticing what needs attention now	Seeing how to prepare for the future
Having patience	Having enthusiasm
Keeping track of essential detail	Watching for new essentials
Facing difficulties with realism	Tackling difficulties with zest
Reminding that the joys of the present are important	Showing that the joys of the future are worth working for

Source: Reproduced by special permission of the publisher, Consulting Psychologists Press, Inc., 577 College Avenue, Palo Alto, CA 94306, from *The Myers-Briggs Type Indicator Manual* by Isabel Briggs Myers © 1962.

where leader behavior is one of the variables. Fiedler's four variables are (1) whether or not the leader enjoys good relations with the team members, (2) whether the task is highly or poorly structured, (3) whether the leader has a high or low position of authority and power, and (4) whether the leader displays a task-orientation or a relationship-orientation toward the group's effort.

Leader-member relations are described as good if the people look forward to working with each other and there is mutual respect and trust. They are described as poor when there is a lack of respect and trust and some people would rather not work with the leader.

A task is said to have high structure when both the goal and methods are well understood and agreed upon and there is

TABLE 5.8
Thinking and Feeling Types Can Supplement Each Other

A Thinking Type Can Help A Feeling Type By	A Feeling Type Can Help A Thinking Type By
Analyzing	Persuading
Organizing	Conciliating
Finding flaws in advance	Forecasting how others will feel
Reforming what needs reforming	Arousing enthusiasm
Holding consistently to a policy	Teaching
Weighing "the law and the evidence"	Selling
Firing people when necessary	Advertising
Standing firm against opposition	Appreciating the thinker

Source: Reproduced by special permission of the publisher, Consulting Psychologists Press, Inc., 577 College Avenue, Palo Alto, CA 94306, from *The Myers-Briggs Type Indicator Manual* by Isabel Briggs Myers © 1962.

little ambiguity about individual roles and responsibilities. A task is poorly structured when any of these ingredients is missing.

A leader's authority and power in the eyes of project team members can arise from having a high position in the organization, from reporting to a very high level on the particular assignment, and from having personal expertise, independent of organizational factors.

Task-orientation refers to focusing on goals, procedures, and assignments, with little attention to how the team members interact. Relationship-orientation, on the other hand, refers to focusing on how the team members behave, with less direct attention to specific goals, procedures, and assignments. In Table 5.9 are shown various combinations of these four variables

TABLE 5.9
Fiedler's Contingency Model of Leadership

SITUATION CHARACTERISTICS								
1. Leader-Member Relations	Good				Poor			
2. Task Structure	High		Low		High		Low	
3. Leader Position Power	Strong	Weak	Strong	Weak	Strong	Weak	Strong	Weak
SITUATION DESIGNATION	1	2	3	4	5	6	7	8
SITUATION RATING*	Favorable			Moderately Favorable		Unfavorable		
PROBABLE LEADER PERFORMANCE								
1. Relation-Motivated Leader	Poor			Good		Good?	??	Poor
2. Task-Motivated Leader	Good			Poor		Poor?	??	Good

*Favorableness of the situation for a leader

together with Fiedler's prediction of how favorable the situation is for the leader and how successful the leader is likely to be. These combinations comprise eight categories numbered 1 through 8, as shown in the table.

By assuming that a leader can be truly successful only when the team is successful, the table can also be read to determine whether a task-orientation or a relationship orientation is more likely to enhance project success. In general, a task-orientation is more likely to enhance project success when the situation is classed as category 1, 2, 3, or 8, and a relationship-orientation is more likely to be helpful when the situation is classed as category 4 or 5 and possibly 6. Whether a task-orientation or a relationship-orientation is better for category 7 is unknown.

A few words of caution are in order before leaving this discussion of Fiedler's contingency model. First, it is one of several contingency theories which in general are concerned with group performance as a function of the situation. Fiedler's model was selected because it seems to relate as well as any to situations like those of a project manager. Second, it is very difficult to replicate study populations and to control experimental conditions in order to validate contingency models. Accordingly, the validity of Fiedler's model, as well as other contingency models, is somewhat controversial.

6

Reporting Techniques

This chapter pertains to reports from the project to its customer, an essential part of assuring customer satisfaction. Every customer who takes a project seriously needs to be appropriately informed about the work done on behalf of the customer. If the customer is either underinformed or overinformed, he or she will not be satisfied, no matter how good the work itself may be.

Underinforming leaves the customer in the dark, wondering if scarce resources are being well-spent. Overinforming floods the customer with material neither needed nor wanted, and it wastes time. Overinforming also conveys the impression that the project staff either does not understand the customer's viewpoint or is unable to distinguish between the important and the trivial.

The need for various reports and their nature, extent, and frequency should be considered when the project is planned. Reporting takes the time of project personnel. The efforts they spend in reporting activities must be in the project plan, schedule, and budget. Most project personnel estimate too little rather than too much time for project reporting when they lay out their activities. The project manager should correct this tendency when working on the overall project plan.

The minimum reporting required of the project may be specified by the contract or scope of work. Even when minimum reporting requirements are defined, however, they may not represent the complete or real needs of those who need to know. Accordingly, the project manager must determine what information really should be transmitted. This often means that the project manager must ask the many-faceted customers about their needs and wants.

This chapter describes six types of reporting mechanisms, considers the selection of a suitable mechanism, gives some tips on preparing reports and making presentations, and discusses the critical role of feedback.

I. REPORTING MECHANISMS

Six ways available for a project to report its work to its customer are:

1. formal written reports
2. informal reports and letters
3. presentations
4. guided tours
5. informal meetings
6. conversations

A. Formal Written Reports

Formal written reports are the mechanism that most project staff, and indeed most customers, think of when they think of project reports. These, together perhaps with presentations (Section C), are likely to be specified in the contract or scope of work.

The virtues of formal written reports are:

1. They are more or less self-contained and complete and are therefore useful to personnel not familiar with the project.
2. They are in a form that facilitates their long-term retention and thereby provide a permanent record of the project.
3. They can accommodate background descriptions, literature reviews, and numerous appendices that might otherwise have no home but are important to understanding various aspects of the project.

Formal written reports also have some liabilities:

1. Because of their potentially widespread and long-term use by people not intimately familiar with the project, considerable care must be taken in their preparation. This requirement makes it difficult for them to be issued quickly or inexpensively.

2. Because the potential audience may be diverse, it may be difficult to satisfy completely the needs of all readers.

Given their virtues and liabilities, formal written reports are often reserved for reporting, analyzing, and interpreting major units of work rather than for reporting progress in real time. Often very little in a formal written report will be new information to the people who are connected with the project; these people will have already learned the essentials through other means. Nevertheless, the project manager should not shirk this activity, for it may be contractually required. Moreover, the reports may carry great weight with those who are not intimately connected with the project but rely on them as their only source of information on what was done and accomplished.

B. Informal Written Reports and Letters

Informal written reports and letters are used to keep the customer up-to-date regarding project progress, accomplishments, difficulties, and near-term plans. They are often prepared on a set schedule, such as daily, weekly, biweekly, monthly, or quarterly. If the regular schedule is biweekly, monthly, or quarterly, supplemental reports may also be submitted when unusually important events occur.

The informal written report or letter differs from the formal report in the audiences that it serves and its ability to stand alone. While the formal report is designed to be self-contained and meaningful even to those who have no prior contact with the project, the informal report is addressed to an audience that is assumed to be up-to-date as of the previous informal report. Thus, the informal report may sometimes dispense with establishing the context, presenting background material, etc. It must, however, refresh the reader's memory if necessary. Therefore, the longer the reporting interval, the more likely it is that some review of the project's previous status or current plan will be needed.

The virtues of informal written reports are:

1. They can be quickly prepared, so they are timely.
2. They are relatively brief, so important items are less likely to be buried and go unrecognized.
3. They can reach a large audience rather inexpensively.

Liabilities of informal written reports are:

1. Lacking full explanations and context, these reports may be misinterpreted by people who have not kept up with the project's progress.
2. As interim reports, some of the information presented may be superseded. However, the readers will not necessarily know if it has been unless they read subsequent reports.

The characteristics of informal written reports obviously make them a very useful mechanism for keeping their readerships up-to-date. They should not, however, be considered substitutes for formal written reports.

Before leaving the subject of informal written reports, we note that some customers expect to see copies of daily field notes or laboratory notes. If this is the case, some attention should be given to how much analysis or interpretation should be sent along. One school of thought says that no analysis or interpretation should be transmitted in this way. The reason is that it may later be repudiated after additional data are obtained or further thought is given to the matter. Another school of thought holds that if the project staff do not offer an explanation, then the customer may apply its own and do so erroneously. Both arguments have their merits. Thus, rather than advocate one school of thought or the other as a general rule, we suggest that the project manager weigh the advantages and risks of each in light of the customer and the situation at hand and act accordingly.

C. Presentations

A presentation is, in effect, the oral equivalent of a formal written report. It is more than just a meeting to discuss project status, although such a discussion may follow the presentation (see Section E). The purpose of a presentation is to take advantage of face-to-face interactions (Chapter 5) while transmitting the significant aspects of a major work unit.

The virtues of a presentation are:

1. An astute presenter can adjust his or her delivery while proceeding in order to reach the audience appropriately.

2. Because a presentation is "live" and not "frozen" as is a written report, it provides an opportunity to try out alternative ways of conveying information and ideas.
3. The captive nature of an audience at a presentation may result in some people paying attention to the project who would otherwise not do so.
4. A presentation affords an opportunity for many different people to discuss a project who otherwise would not be current on its status and take the time to do so.
5. An audience may understand information to be conveyed by a graph, table, or chart better when it is presented than when it is written.

Liabilities of a presentation are:

1. The presenter must be able to retain poise when questioned by members of the audience.
2. If the presenter is not expert in all facets of the work being reported, it may be necessary to have several experts attend the presentation.
3. A presentation is not easily made to a geographically dispersed audience.
4. A presentation does not *per se* produce a permanent record.

It is often appropriate to use both a presentation and a written report. This approach offers the advantages of both and helps overcome some of their disadvantages.

If the presentation precedes the publication of the final report, it can be used to test ideas before they are printed or to introduce the audience gently to material they might otherwise find surprising or distressing. Also, a discussion following the oral presentation may provide suggestions that can improve the written report.

D. Guided Tours

"Guided tours" refers to taking customers to the project site in order to familiarize them with the work. Sometimes there is no good substitute for a first hand examination of the project situation. Yet many a customer has never visited its project site. Accordingly, such a customer has little appreciation of the project manager's situation. In these cases, the project manager will do both the project and the customer a favor by organizing a site visit.

Virtues of site visits or guided tours are:

1. They develop insight on the part of the customers which may be difficult if not impossible to develop any other way.
2. They tend to be informal, which makes it easier for the participants to get to know each other and thereby develop trust.
3. They tend to be absorbing, which means that the customers' attention will less likely be divided.
4. They tend to take a fair amount of time, which provides multiple opportunities to communicate.

Liabilities of guided tours are:

1. They are difficult to arrange if many people must be coordinated or if they are remotely located from the normal activities of the individuals.
2. They are relatively expensive to conduct, especially if distances are great.
3. They may interfere with the normal activities at the project site.

While the direct and indirect costs of a site visit are not insignificant, there are times when the visit is perhaps the only effective way to report to the customer the real essence of the project. If this is the case, the project manager should include one or more site visits in the project plan, schedule, and budget. These visits will pay the largest dividends if at least the first one occurs rather early in the project.

E. Informal Meetings

Informal meetings are often the most productive way to report to a customer. While overall they may not satisfy contractual stipulations, they are more likely to produce a two-way report, i.e., from the customer as well as to the customer. As mentioned in Chapter 2, information from the customer constitutes a vital input for the project manager.

Informal meetings may either follow a presentation or be called separately. Their virtues are:

1. They are in fact informal, so that the formalities that accompany a presentation are usually missing. This enables the attendees to use their time to good advantage if they practice good meeting principles (Chapter 5).
2. The participants may be but a subset of those who might attend a presentation, so that the discussion can be limited to the needs of a limited group.
3. They generally provide a more inviting atmosphere than a presentation, so all the attendees feel able to participate, not just a few key people.
4. They can usually be arranged with relatively little fuss and only moderate advance notice.

The main liability of informal meetings is that the right people might not all be present, so some decisions may have to be reviewed or reconsidered before they can be implemented.

Needless to say, the principles of good meetings discussed in Chapter 5 for project staff meetings apply to meetings between the project and its customer. There may be a slight problem, however, in deciding who is in charge of the meetings.

When informal meetings between the project and its customer are called at the request of the project manager, the customer's chief delegate may nevertheless want to be in charge. If the latter chairs the discussion, the project manager may or may not get the issues resolved satisfactorily, depending upon the skill of the chair. If the project manager suspects that the chair will not be effective, he or she may try to arrange in advance to lead the discussion once the introductory remarks have been made. The chair can then summarize and conclude the meeting after the discussion is over.

F. Conversations

Conversations have the same advantages in reporting to the customer that they have within the project organization (Chapter 5). They are, however, hardly ever recognized as an official vehicle for the project to report to its customer. They are simply too casual to satisfy inter-organizational responsibilities. This does not mean that they cannot or should not be used.

Rather, conversations should be used between the project and its customer to establish understandings. Once these understandings have been reached, they should be confirmed by an informal report or a letter (see Section B). In this way, the project manager can have the benefits of conversations and still practice good reporting technique.

II. CHOOSING A SUITABLE REPORTING APPROACH

In choosing a reporting approach, the project manager should consider the nature of the audience and their needs and the time and resources available. The audience can be characterized along five dimensions:

1. *Audience diversity* It is difficult to serve a widely diverse audience with a presentation. One part of the group may be bored while the needs of another part are being addressed. The other approaches listed in Section I can all accommodate a diverse audience if suitable care is exercised.
2. *Audience sophistication* An unsophisticated audience may benefit most from an informal approach--whether in writing or orally—that borders on the tutorial. And an oral approach, with two-way interaction, is generally more effective in this case than an informal written report. A more sophisticated audience may prefer a more formal report or presentation which is then followed by discussion.
3. *Audience familiarity* If the audience is familiar with not only the project as described in the scope of work, etc., but also the project site, then written and oral reports and presentations will probably be effective. On the other hand, if the audience is not familiar with the project site, then a guided tour is in order.
4. *Audience size and geographic location* If the audience is large or widely dispersed, then written reports may be the only viable means. But if it is compact in both size and location, then presentations and informal meetings should be considered for the two-way benefits that they offer.
5. *Audience's need to know* Do the members of the audience all have the same needs to know? If so, presentations and informal meetings may be effective. If not, written reports are probably better providing they make clear where each segment of the audience will find what it needs.

Project managers will probably feel constrained by the schedule, budget, or availability of key personnel as they try to optimize their reports to clients. They need, therefore, to consider ways of compromising that will not at the same time seriously harm their projects.

They should consolidate reporting efforts where practical in order to minimize redundant work. They should schedule reporting activities at opportune times, not necessarily according to rigid schedules, so that project staff can both perform their work and report on it with a minimum of extraordinary stress. And they should arrange reporting events, especially presentations, well enough in advance that the reports are timely when received. This latter practice will minimize the need for *ad hoc* reports to selected individuals who would otherwise request or require special reports.

III. PREPARING REPORTS AND MAKING PRESENTATIONS

This book is neither a text on expository writing nor a treatise on public speaking. Yet it seems in order to present a few tips on effective reports and presentations that are frequently overlooked in the heat of project activity. These few suggestions together with some basic communication skills will help project managers report successfully to their customers.

1. Plan the report or presentation. What are the objectives for the communication? What does the audience already know? How can one get the audience from their present levels of understanding to the objectives? First, the route should be outlined. Then, the details should be filled in.
2. Capture the audience's interest at the outset; then develop the subject. If the subject is developed before their interest is captured, they may not pay sufficient attention to get the message.
3. If preparing a written report, be sure that all figures can be interpreted without reference to the text. Also, figures should not require color reproduction to be interpreted, since many figures are later photocopied in black and white.
4. If making a presentation, be sure that the physical conditions are suitable. Otherwise, they will distract the

audience. The air temperature and quality, noise and lighting levels, the quality of seating, and the audience's ability to see any audio-visual aids and to hear the speaker all deserve attention.

5. When making a presentation, practice good delivery techniques:
   a. Define and post the objectives for the presentation and tell the audience why they are important. Do not assume that the audience knows either the purpose or why it is significant.
   b. Tell the audience what the path or topic sequence will be. They will receive the elements of the delivery serially. Since they cannot ruffle through the delivery like a book to see where various elements come, they need to be told in advance in order to put the pieces in place as the presentation proceeds.
   c. Use visual aids as appropriate. They should help the audience follow and understand the presentation but they should not be a substitute for it. That is, the audience should get more from the total presentation than they get from the visual aids alone. See also item 6 below.
   d. Build bridges from one topic to the next to help the audience follow the presentation.
   e. Do not give needless details. Many members of an audience are offended when they receive more information than they need.
   f. Sum up major points so that the audience will recognize them.
   g. Recommend action when it is appropriate. Most audiences need a concrete proposal to focus on when they contemplate future action.
   h. Watch the audience's reaction to the presentation and adjust it to get desired results. Avoid a stilted delivery that puts the audience to sleep. Establish and keep credibility by being candid and forthright. (And keep enough lights on in the room to be able to read the expressions and other non-vocal cues of the audience.)
6. When using visual aids, practice good techniques:
   a. Use lettering and symbols large enough and in sufficient contrast with the background that they can be read from the back of the room. Demonstrate that they are satisfactory *before* the presentation and make corrections as necessary.

b. Put related items on the same chart if doing so does not result in crowding the chart. Do not distract the audience with unrelated material. Sometimes cover a list on a chart and expose each new item only when it is time to discuss it. This technique is helpful when the list is long and the time devoted to each item is also significant. The technique keeps the audience from reading ahead to later items while an earlier item is still being discussed.
c. Do not merely read a chart to the audience. Explain or amplify it. Give examples. Otherwise, the presentation can be reduced to a silent showing of charts. At the same time, use the same kind of language as the charts. Do not force the audience to translate one or the other to see their equivalence. The object is to have the visual aid reinforce the message and increase audience understanding, not increase audience confusion.
d. Provide copies of the charts for the audience to use in note-taking.

IV. OBTAINING FEEDBACK

It pays project managers to obtain feedback from those who receive their reports and presentations. Feedback enables project managers to determine if their materials were understood and provides an opportunity to correct misunderstandings if they are detected. Feedback often includes information that is helpful for future action and enhances project managers' abilities to satisfy their customers. And requesting feedback helps project managers show that they care whether their audiences understand.

Feedback can be obtained by a personal approach, i.e., interviewing; by questionnaire; or by both. If a personal approach is used, the project manager can interview members of the audience "cold" or can tell them that he or she will phone or visit at a particular time to discuss their reaction to the report. The latter may also defuse any need members of the audience have to send an angry letter.

A questionnaire enables the project manager to reach a larger audience, but it requires very careful preparation if it is to elicit useful information and also not offend the respondents. In fact, most questionnaires need to be "dry run," perhaps several times, to make them unambiguous, efficient, and helpful.

They are generally used therefore when so many people must be surveyed that the cost of preparing a suitable questionnaire is less than the cost that would be incurred if the same people were to be interviewed.

When a questionnaire is in order, it may also be supplemented by interviews of selected individuals to learn about their special reactions and needs. These individuals may be people who have a special relationship to the project or people who raise special points in their replies to the questionnaire. The latter of course, may be detected only if the questionnaire solicits comments in addition to asking the respondents to choose from predetermined answers.

In addition to obtaining feedback from the audience of a report or presentation, the project manager can also query the project staff. They, too, need to be shown that the project manager cares what they think, and they may have useful information that they would not volunteer without being asked.

7

Negotiating

To negotiate is to confer with another so as to settle some matter. Project managers frequently must negotiate with clients, bosses, functional group managers, vendors and subcontractors, support group managers, and their own project staff in order to resolve issues such as scope of work, standard of work performance, staff assignments, schedule, budget, and so forth. Negotiating is so pervasive in project work that the project manager needs to know how to negotiate successfully. This chapter, therefore, describes the elements of negotiating, how to prepare for a negotiation, arrangements for negotiating, and negotiating techniques.

I. ELEMENTS OF NEGOTIATING

A. Cooperation

Negotiating is a cooperative enterprise in which two people or parties search for an arrangement that leaves both of them better off than they were when they started. Essentially, each party makes concessions in such a way that the concessions that it receives are worth more *to it* than the concessions it gives up to the other party. The aim is to find a "win-win" arrangement and to avoid a "win-lose" arrangement. Searching for a "win-win" arrangement indeed requires cooperation, and negotiating is thus a cooperative, not a competitive process.

B. Something for Everyone

Each party in a successful negotiation has to let the other party win something of value to it. Otherwise, that party has no reason to participate and may withdraw from the negotiation. If one party withdraws, then neither party can win anything. That is, concessions cannot be obtained from a party who refuses to negotiate.

If one party enters a negotiation with the idea that it will utterly vanquish the other party, it is not psychologically prepared to negotiate. As soon as the other party discovers that it cannot win something of value, it can and should terminate the negotiation. Neither party should continue to negotiate, i.e., continue to trade concessions, if it is receiving less than it is giving. Thus, each party must be prepared to let the other win something.

C. Satisfaction of Real Needs

A successful negotiation satisfies the real needs of the two parties. A real need is one that is considered genuine by the party who is presumed to have the need, not the party who is offering the concession.

To put the matter another way, there are rewards and non-rewards. A non-reward is not a punishment; it is just not a reward. Something which the other party does not value is a non-reward, no matter how valuable or rewarding the offering party thinks it should be. Non-rewards do not satisfy needs, whereas rewards do. Each party must search for rewards that the other party will value if it hopes to negotiate successfully.

While a party can offer a non-reward, it will not advance the negotiation nor induce any concessions from the other party. It may, however, be accepted by the party to which it is offered. If so, the offering party gives up something needlessly and gets nothing in return. Overall, it may be worse off for having offered the non-reward.

A key skill of a successful negotiator is the ability to determine quickly and with some accuracy the needs of the other party. One thus avoids offering schedule relief when the other party truly needs more budget. One does not offer extra financial gain when prestige is the real issue. And one does not offer security when self-actualization is the other's objective. By the same token, one does offer schedule relief when possible *if* it meets the other party's need. And so forth.

In assessing another's needs, it is risky to impose one's own standards. We have all heard of artists who live happily in hovels while intently pursuing their art for little or no pay. We might believe that their living conditions indicate a need for better housing. This is not necessarily true, and an offer of better housing could be a non-reward for them. Simple, even primitive shelter can satisfy their needs while it might not satisfy ours.

The other party's needs and constraints can be inferred from a variety of sources:

1. budgets and financial plans
2. publications and reports
3. press releases
4. officers' speeches and public statements
5. institutional advertising
6. reports to agencies, e.g., the Securities and Exchange Commission
7. Moody's and Standard and Poor's reports
8. recorded real estate deeds
9. credit reports
10. instructional and educational materials
11. the other party's library
12. biographical references, e.g., *Who's Who*
13. others who have faced the same party

The sources may suggest rather concrete needs, such as the need to complete a project by a particular date or the need to accomplish certain work at the lowest possible cost. But some sources may also suggest less tangible objectives, such as being master of one's own affairs, being regarded as an influential person, or being protected from excessive risk.

D. Common Interests

Negotiating involves seeking common interests. The two parties need to identify together what each wants and what each can give up. It is best if such information is traded in small amounts; this approach helps keep a cooperative spirit going until agreement is reached. Also, trading information in small amounts helps each party avoid offering a substantial concession without receiving a commensurate concession in return.

E. A Behavioral Process

Negotiating involves trading information and concessions in a somewhat formal way. It is a little like a minuet, with each party behaving in a way that is understood by the other. This behavior enables both parties to estimate about how much information or concession to offer at any moment and how much to expect in return. Each party thus feels that the concessions that they are exchanging approximately balance. Each is therefore willing to continue the negotiation until both are satisfied with their overall situations.

II. PREPARING FOR A NEGOTIATION

A project manager can generally enhance the chance of a successful negotiation if he or she prepares appropriately for the event. This section describes several steps that can improve the likelihood of a satisfactory outcome.

A. Determine Objectives

The project manager should determine the objectives for the negotiation and should know which aspects of the desired results are absolutely essential and which are merely nice to have.

Some assessment should also be made about the difficulties to be expected in trying to obtain the various objectives. This assessment will be useful in establishing an agenda for the negotiation (see paragraph D).

B. Determine What Can Be Yielded

Of those items that are wanted, the project manager should determine which can be given up if doing so would meet a need of the other party. They should be ranked in priority so that the "less expensive" ones can be offered before the "more expensive" ones (as seen by the offeror). This will enable the negotiator to offer items of small value to it in order to induce concessions and build goodwill in the early stages of the negotiation (see paragraph D and Section IV).

Also, each party should determine what cannot be yielded. It needs to avoid inadvertently yielding an element that it considers essential.

C. Determine the Other Party's Real Needs and Constraints

Each party should understand the other party's real needs so that it can try to meet them. It should also understand the other's real constraints so that it does not make demands that cannot possibly be met. If the other party is aware that it is being asked to violate its own real constraints, it may withdraw from the negotiation, for it is bound to lose if it stays in.

D. Prepare a Tentative Agenda for the Negotiation

The project manager should prepare a tentative agenda for the negotiation. Since the other party should also prepare a tentative agenda, the first item on the agenda, after introductions and preliminary remarks, should be to agree on the actual agenda that will be followed.

Generally speaking, the agenda should progress from minor items and items that will be easy to agree upon to major items and items that may be difficult to agree upon. The point here is that each party needs to develop a feeling of cooperation and goodwill toward the other before attempting to negotiate major and difficult issues. This is most readily done when small concessions are exchanged in the early stages. Then, by the time the difficult issues are reached there will be a history and a spirit of satisfying mutual interests. This experience and attitude will help the two parties accommodate each other where possible in order to reach a mutually satisfactory conclusion.

E. Determine Your Image in the Other Party's Eyes

A party to a negotiation should try to determine how it is seen by the other party. Suitable forewarning can help it prepare to combat unproductive behavior by the other side.

For example, if the other party thinks that it is facing a pushover, it may make extraordinary demands. Then, it may feel "honor bound" to obtain these demands even if it later learns that they cannot be granted. The situation becomes a matter of saving face for the party making the demands and it can stymie the negotiation. This type of impasse can be removed by offering an alternative approach instead of just denying the requested concession. Or, occasionally the request can be broken into several parts and some of them given up.

If the other party thinks that it will be dealing with a very tough negotiator, it may behave defensively and be reluctant to exchange information that is vital to concluding a successful negotiation. In this case, it pays to treat the other party warmly and to build its confidence in one's own reasonableness.

If one expects the other party to err in assessing its opposition, then it is well to try to revise the other's impression from the beginning. This can be done when the negotiation is being arranged and when the agenda is being determined. One can, for example, show that he or she is reasonable to deal with by accommodating the other party on issues that have little significance. At the same time, one can display the ability to stand steadfast to principles when there are in fact important positions to protect.

F. Develop Alternative Solutions

A party about to enter a negotiation should develop alternative acceptable solutions to the difficult issues, in case its first approach is unsatisfactory to the other party. Since common interests are to be sought (paragraph I.D), some exploration of alternatives is in order during the negotiation. The proceedings can be facilitated if each party knows in advance some alternatives that are acceptable to it and those that are not. Such preparation will enable it to suggest or focus attention on acceptable solutions and to steer away from unacceptable solutions.

G. Decide Who Should Negotiate and Assign Roles

There are two aspects to selecting a negotiator for a particular negotiation. One aspect is to decide whether the negotiator should be an individual or a team, and the other is to decide who it or they should be. If a team is chosen, there is the further aspect of assigning roles to the team members.

A single negotiator has certain advantages and disadvantages:

1. It avoids questions being asked of weaker members, since there are none, and thereby avoids having their answers undermine the work of the team leader.
2. It avoids disagreement among team members which the other party could use to attack the leader's arguments.
3. It facilitates on-the-spot decisions to make or gain necessary concessions.
4. It places complete responsibility in one person, except as he or she is restricted in authority and therefore required to seek the approval of others who are not present at the negotiation.

Team negotiation also has some advantages and disadvantages:

1. It can provide people with complementary expertise to correct misstatements of fact.
2. It provides witnesses as to what was said.
3. It enhances the reading of the other party's body language by having more observers.
4. It enables the pooling of judgments.
5. It presents the other side with a larger opposition.
6. It requires one person to be designated the chief negotiator, who serves as the focal point for the team, directs the participation of individual team members, and makes on-the-spot decisions if necessary.

Generally speaking, if the members of a negotiating team are well-instructed and well-disciplined, the team approach is the better of the two when major issues are involved or the outcome is crucial. Even a two-person team can be much more effective than a single negotiator. A team approach assumes, of course, that the team members are available for the duration of the negotiation.

A single negotiator may be better when the issues are relatively minor; when they must be resolved promptly, i.e., without making formal arrangements involving several people; or when a "public" exercise might cause one party or the other to behave unproductively in order to be consistent with previous positions or otherwise save face.

An important factor in selecting a negotiator or negotiating team is the relative position of the other party's negotiator(s). Contrary to a commonly held view, one should not send a negotiator or team that has more authority than the other party's negotiator(s). Since negotiation is largely a matter of obtaining and making concessions and commitments, one does not want to be in the position of being bound by one's own commitments while the other party's negotiator(s) can have theirs reneged. If this happened, one would almost certainly end up giving much more than one receives. The argument applies to the other party as well. Thus, in a well-planned negotiation, the two parties will send negotiators whose authorities are approximately the same in terms of the kinds of commitments that they can make.

In the event that one does send a negotiator whose authority is significantly greater than that of the other party's negotiator(s), the negotiator with "excess authority" can simply indicate that he or she must have the subsequent approval of others (who might be named) before proposals, etc., become commitments.

The need to designate a chief negotiator in a team negotiation was mentioned above. When a team is used, other roles should also be assigned. If issues of different types are involved, e.g., technical, financial, liability, schedule, etc., then someone should be designated to negotiate each type. This does not mean that one person cannot or should not do them all. It merely means that the entire team should know who is going to do each type.

Those who are not carrying an active role at the moment should observe the other party's negotiators. The aim is to determine from spoken remarks and from body language who is responding favorably, neutrally, or unfavorably to the concessions and commitments being proposed. This information can then be exchanged among the team members during caucuses in order to devise successful strategies. Each person of the other party should have a designated observer from one's own team. Some team members may of course have to observe more than one person.

A team member should also be designated as note-taker for the team in order to capture details that may otherwise escape recall. The note-taker needs to understand the essence and nuances of the negotiation in order to pick up key points. It also helps if he or she can glean essential information from simultaneous conversations in the event more than one person talks at once. Observational notes and other ideas should also be jotted down by the other team members as they occur so that they can be referred to and discussed in caucuses.

Finally, assignment of roles includes instructing the team members on how to behave when they are and are not the lead negotiator at the moment. The lead negotiator needs not only to conduct the discussion at the time but also to be on the lookout for signs from team members that they need to communicate with the leader before he or she proceeds. The leader should not go so fast so as to preclude receiving their inputs or force them to interrupt in order to steer the discussion away from unattractive areas.

By the same token, the other team members need to indicate when they have an input to make first to the team leader and then perhaps to the entire group. They might do this by a signal of the hand or eye, or they might pass a note. Sometimes they might ask for a moment to confer privately with the leader. Generally, they should try to get the leader's attention without interrupting. There are times, however, when the leader seems oblivious to the more subtle approaches and an interruption is necessary.

A team member should not risk speaking directly to the other party without first getting an okay from the leader at the moment unless there has been prior agreement that he or she is the spokesperson on the subject or aspect at hand. If there is any uncertainty about authority on the matter, he or she should seek the leader's approval before speaking.

Team negotiations require exceptional teamwork. As in rowing, no one is praised for rugged individuality.

H. Practice

Practice is recommended before a serious negotiation for two reasons:

1. It helps identify the difficulties that will be faced in the negotiation so that alternative approaches can be developed and evaluated.
2. It permits the would-be negotiators to rehearse their roles and develop some skill in exercising them.

Obviously, the more skillful and knowledgeable the negotiators, the less practice they need. But even the most experienced negotiators generally profit from at least one "dry run." A "dry run" allows them to heighten their awareness of the issues and, if it is a team negotiation, to improve their coordination and ability to work together.

An effective way to practice is to have personnel who will not be part of the negotiating effort role-play the other party in a mock negotiation. If an experienced negotiator can be enlisted as a critic and coach during the mock negotiation, so much the better.

In general, the amount of preparation needed for a negotiation depends upon several factors:

1. the skill of the negotiator(s)
2. the relative importance of the outcome
3. the alternatives available

Many negotiations are conducted on the spur of the moment, with virtually no obvious preparation. However, chance favors the prepared mind. If project managers are aware of and pay attention to the issues discussed above as they proceed, they will enhance their results compared to spur-of-the-moment negotiations. For more formal arrangements, they can and should take the trouble to prepare more formally.

III. ARRANGEMENTS FOR NEGOTIATING

The arrangements for a negotiation should be chosen with the idea that they can help attain one's negotiating objectives. The variables are the time, the place, and the schedule for the negotiation.

A. Time

"Time" refers to when negotiating should begin. Adequate opportunity should be allowed for both parties to prepare for the negotiation. Attention should be given to when the parties are available to negotiate. And the negotiation should be conducted and concluded in time for the outcome to be useful to the project.

In many cases, preparation, availability, and time give conflicting answers about when negotiating should begin. Also, the two parties may have different needs, so they may have conflicting ideas about when to begin. When either or both of these conflicts exists, a compromise is in order.

A compromise on the starting time should not be accepted, however, if it would disadvantage the accepting party to the point that it would be better off if it did not negotiate at all. It requires courage to decline to negotiate because the terms under which it would have to be done are unacceptable. But discretion is often the better part of valor, and that is true here.

B. Place

The selection of the place for a negotiation involves such factors as freedom from interruptions; psychological advantage; and access to personnel, equipment, files, and services. The first factor is self-explanatory, but the latter two warrant some discussion.

Some psychological advantage generally accrues to the party that negotiates on its own premises. For this reason, most negotiators like to work at their own place of work or business. Since this is true of both parties, they may choose to favor neither party and instead hold the negotiation at a neutral site.

There are times when the negotiator should offer to negotiate at the other party's place of work or business. For example, if the other party would be less defensive and more cooperative if it felt more at home, then it might pay to offer to negotiate at its place. And if it would serve to build good will by meeting at the other party's place, then the negotiator may offer to do so. In either case, however, a negotiator should not make such an offer if doing so would create a serious disadvantage.

Sometimes it is helpful if a negotiator can control access to other personnel and files during the negotiation. One may not want to have certain experts or information readily available if finessing the issues would give an advantage. There are times when the other party will not pursue an unattractive element if it is not available or will give the benefit of the doubt if the facts cannot be easily obtained. For this reason, a negotiator may prefer not to negotiate on his or her own premises.

On the other hand, a negotiator may choose to negotiate where selected personnel, equipment, or files are available because they are needed to make certain arguments and they cannot easily be taken to another site. Likewise, a negotiator may prefer to negotiate on his or her own premises because it has conference telephone call facilities, stenographic services, or even just a pleasant atmosphere that may enhance or expedite the proceedings.

C. Schedule

"Schedule" refers to the pacing of the negotiation. Generally, whenever a party feels some urgency about concluding the negotiation, it is likely to make concessions more readily than it otherwise would. It pays therefore not to be the party with the earlier deadline for concluding the negotiation.

Sometimes a party would like very much to conclude a negotiation by a particular date or time. However, it should not reveal this fact. While it can indicate its hopes if asked, it should make very clear that it is able to take as long as necessary to work out a mutually satisfactory agreement. Otherwise, the other, more relaxed party may stall until the deadline approaches. As the deadline nears, the more anxious party may lose patience and become unwilling to work for what it wants. It can end up giving concessions more easily and demanding less in return than it would if it were not so anxious.

Even when one party does not press the other against a deadline, the negotiation may need to be concluded promptly for the result to be timely. When this happens, one party (or both) may be unable to strike the bargain that it would like to have. It should not be disappointed, however. When half a loaf is better than none, the correct comparison is between something and nothing, not between something and everything.

By the same token, a party should not let schedule deadlines force it to negotiate a truly unsatisfactory result. It should instead withdraw from the negotiation. As mentioned in Section I.B, if one cannot be better off after the negotiation than before, there is no reason to participate.

IV. NEGOTIATING TECHNIQUES

The term "negotiating techniques" has two meanings. The first meaning refers to practices or techniques that help negotiators exchange information and concessions and help them avoid making unwanted concessions. The second meaning refers to psychological pressures that one party may apply to the other in order to obtain concessions that the latter would not make in the absence of the pressures. Techniques of the first type are discussed in this section. Information on techniques of the second type can be found in the popular literature.

A. Defining the Issues

It is important for the negotiators to define the issues to be negotiated. Each party should name its concerns and propose a sequence or an agenda for considering the issues. The two parties should agree upon a common agenda before proceeding to the issues themselves.

As discussed in Section II.D above and Section IV.C below, it is generally helpful if the issues that appear difficult to either party are placed toward the end of the agenda. Thus, each party should let the other defer whatever issues it wants. If this is done, only issues that seem to both parties to be minor or easily resolved will come first. As the easy issues are resolved, both sides will build a reservoir of good will toward each other that will help them resolve the more difficult issues.

If one party wants to negotiate an issue that the other feels is non-negotiable, it is best to put the issue on the "difficult" list and defer its consideration until after the other issues have been resolved. Perhaps the second party will change its mind if everything else goes well. Even if it does not change its mind, it can wait until the issue is about to be actively discussed to declare that it is non-negotiable.

B. Establishing the Participants' Authorities

Since negotiating consists essentially of exchanging concessions and each party wants to receive at least as much as it gives, it is important for the two parties to have commensurate authority to make binding concessions. Neither party should be empowered to give greater concessions that it can receive from the other. Otherwise, the party with the greater authority will find itself making significant concessions that it must honor while the other party will have to honor only less significant concessions.

To avoid making more significant concessions than it can receive, each party must know the authority of the other. If one party determines that it in fact has more authority than the other, it can shed some of its authority for the time being by declaring a need to consult with absent personnel (see Section D).

If one party does not know the other's authority to make binding concessions, it should ask. Simple, straightforward questions are best, such as "To what level can you make a commitment without the approval of others?" or "Are there any areas that we are about to discuss that must be ratified before our agreements are binding?"

C. Building Goodwill

Goodwill facilitates the negotiating process. At some point, one of the parties has to offer a concession in order to induce the other to give one in return. There is no assurance, however, that the offer will not be rejected. The very prospect of rejection can inhibit the first party from making the offer. This inhibition is diminished markedly, though, when there is a feeling of goodwill that keeps a potential rejection from being taken personally.

Goodwill is developed by the parties behaving in a caring way toward each other. This can occur naturally as each party accommodates the other on details of the arrangements that are not of great importance to it.

Goodwill is also developed by resolving some issues easily. When agreement is reached, the two parties usually feel good about their accord, for it reduces some of the tension and anxiety that accompanies all negotiations.

As a reservoir of goodwill is developed, it is easier for the parties to cope with the tensions that accompany the more difficult issues. Indeed, when the goodwill is adequate, it may be as easy to deal with a difficult issue as it was to deal with an easy issue before there was significant goodwill.

D. Seeking Approvals of Absent Personnel

The approval of absent personnel is sought when a negotiating party either truly lacks authority or wishes to avoid having authority to make binding commitments on-the-spot (see Section B). All commitments that are made subject to another's approval are, of course, subject to being reversed.

If one finds that the other party needs to have its tentative "commitments" confirmed by an absent person before they become binding, then one should make one's own "commitments" conditional upon the confirmation. If the confirmation is not forthcoming, then one is relieved of the conditional "commitments."

E. Seeking Higher Authority

Occasionally one party finds the other intractable and wants to deal with a higher authority. If so, the first party should ask to talk with the other party's superior. The request is likely to be granted.

If the other party asks to talk with one's own higher authority, the higher authority should be briefed thoroughly before entering the negotiation. Moreover, he or she should not be allowed to negotiate alone but should be accompanied at all times by someone who has attended the negotiations up to that point. This will avoid the other party's trying to renegotiate old issues.

F. Interrupting Unattractive Lines of Pursuit

Sometimes a party wants to avoid discussing a presently favorable but indefensible position. When this happens, the first party wants to interrupt the other party's line of pursuit and do so with a minimum of fanfare. This will minimize the chance that the issue will be raised again.

Possible means of interrupting are to call for a coffee break, ask for a caucus, adjourn for lunch, distract the conversation, go to the rest room, etc. Upon resuming the meeting, one can begin with another subject, thereby making it difficult to reintroduce the troublesome topic.

When there is a team negotiation, the chief negotiator should warn colleagues about the possibility of having to interrupt unattractive lines of pursuit. Otherwise, one of them may unwittingly continue or reintroduce an unwanted issue in an attempt to tie up loose ends.

G. Securing Commitments

Both parties to a negotiation need to confirm in writing what they have agreed to. This agreement should be done on the spot at least session by session if not issue by issue. These written agreements prevent reneging and arguments about the results of earlier discussions. They thus facilitate forward progress of the negotiation.

If no other means is conveniently available, the agreements can be captured on a flip chart or just written long-hand on paper. Both parties can signify their agreements with the written versions by initialing them.

Epilogue

Project management is an integrating activity and is done best when the manager has a holistic outlook. A typical project involves numerous conflicting pressures and requires artful compromises. These compromises can be made successfully only when the project manager understands how all the elements of the project interact and work together to accomplish the project's objectives.

For ease of explanation and understanding, many of the chapters in this book deal with individual aspects of project management. This is not to imply, however, that these aspects stand alone or that they can be pursued independently. For example, one usually cannot plan without some negotiation, and one cannot negotiate without some prior planning. One cannot control without planning, monitoring, and directing and coordinating. Nor can one report without monitoring. Thus, we urge our readers to blend the several ingredients thoroughly in applying them to their projects.

Many projects get in trouble, and some become utter failures, i.e., time and money are spent but the objectives are not achieved. While the reasons claimed for trouble or failure are varied, they generally boil down to inadequate project management.

Sometimes project managers accept assignments under impossible conditions, perhaps by overselling themselves or buying bills of goods. Competent project managers will not deliberately work with such odds, and they structure their objectives and preliminary studies to protect against ultimate failure.

Sometimes project managers fail to exercise their prerogatives. They may, for instance, accept changes in scope without getting prior approval from their customers for the consequential changes in budget or schedule; later they are late and overbudget and may be unable to collect for the extra efforts. Or they may not ask for timely, meaningful reports from their task leaders; then they are unable to coordinate their work. Or they may hide their problems from management, who are therefore unaware of their special need for resources.

Sometimes project managers dupe themselves into thinking that they are charismatic or omnipotent and able to surmount any shortcoming, whether poor estimates, sloppy plans, inadequate contingency allowances, no controls, schedule compression due to late starts, or ineffective communications. Only toward the ends of their projects do they realize that their powers are finite and quite likely too little to rectify the shortcomings. It is perhaps sad to see project managers lose their naivete and discover that they are not omnipotent. But the true tragedy is the waste of their project teams' efforts and their customers' resources. They are people of good intent and good will, and they have labored in vain.

Project managers are in a position of trust. They are not only the focal points for their projects, they are also the stewards of all the talent and resources committed to the projects. They have a noble responsibility.

Successful project management is a work of art and a joy to behold. A project failure can cause much tragedy. We hope that this book will help you manage your projects successfully and give you, your project teams, and your customers much happiness and satisfaction.

Appendix

Cost Versus Time Profiles

The cost versus time profile of most, but not all, projects follows an S-curve, such as the one in Figure A-1. The various parts of the curve are identified by number and have the following characteristics:

Stage 1. Stage 1 is the start-up portion of the project, while the project manager plans, negotiates for staff, and generally sets things up. Costs accumulate slowly during this time unless expensive equipment is purchased.

Stage 2. Stage 2 represents a rapid expenditure of funds as work proceeds at a fast clip. A cost control system that cannot keep up with the expenditures at this time leaves the project manager vulnerable to missed objectives with inadequate resources to recoup.

Stage 3. Stage 3 corresponds to the time when most of the work is done and attention is devoted to report writing, tying up loose ends, etc. Expenditures are made at a lower rate again.

Stage 4. Stage 4 (or point 4) refers to the time when all explicit commitments have been met. Expenditures to this date should not equal the total budget because there are still post-accomplishment activities to be closed out and charged against the project budget.

Stage 5. Stage 5 is the period when the project is finally closed out and minor last costs dribble in. These last costs plus those incurred in accomplishing the project through Stage 4 should not exceed the overall project budget as finally negotiated and approved.

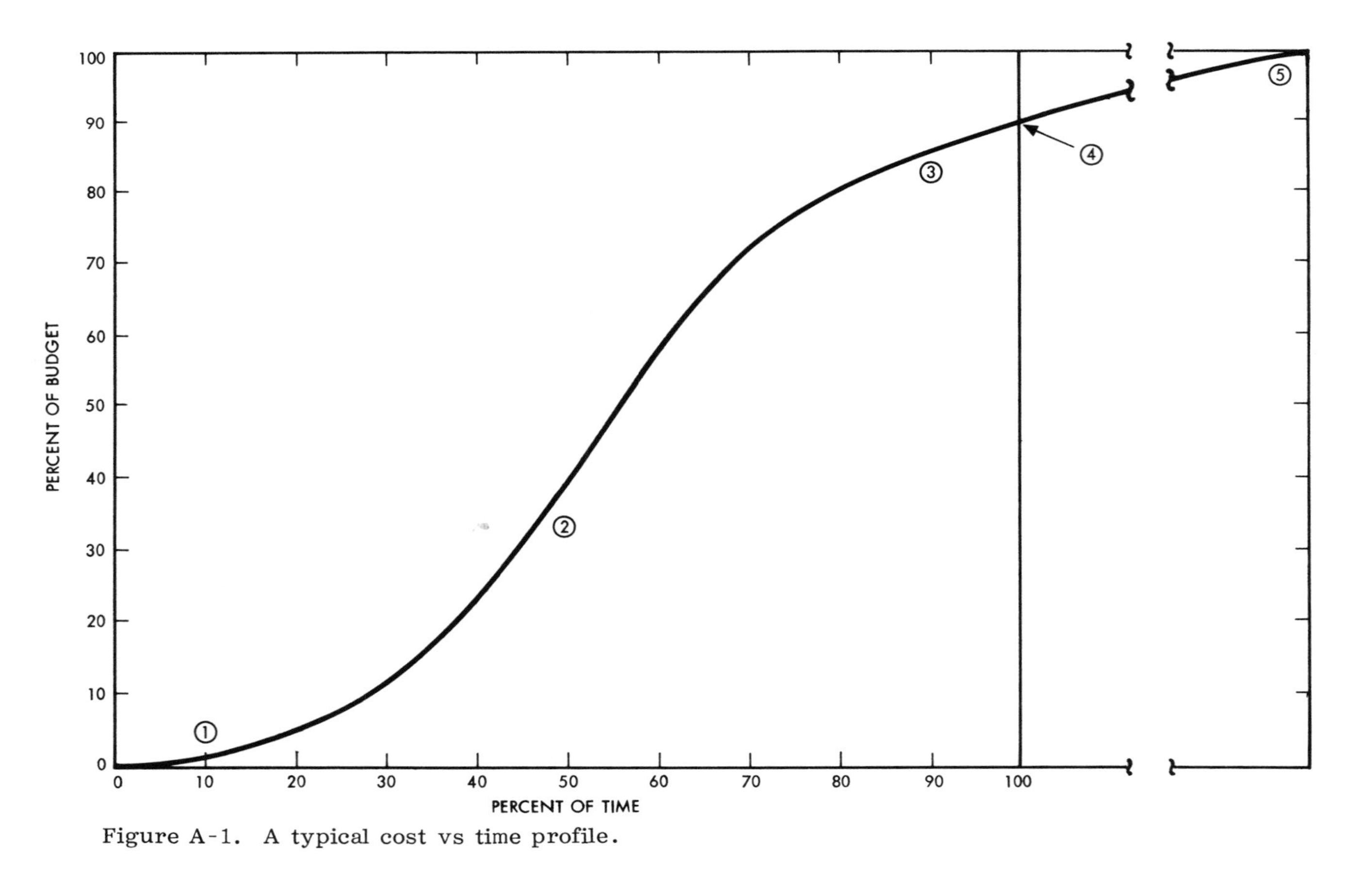

Figure A-1. A typical cost vs time profile.

Bibliography

Abramson, B. N., and Kennedy, R. D., MANAGING SMALL PROJECTS, Redondo Beach, California: TRW Systems Group (1969) 68 pp.

Archibald, R. D., MANAGING HIGH TECHNOLOGY PROGRAMS AND PROJECTS, New York: Wiley-Interscience (1976) x + 278 pp.

Avots, I., "Why Does Project Management Fail?," CALIFORNIA MANAGEMENT REVIEW, *12*, No. 1 (Fall 1969) 77-82.

Baker, B. N., and Wilemon, D. L., "A Summary of Major Research Findings Regarding the Human Element in Project Management," PROJECT MANAGEMENT QUARTERLY, *8*, No. 1 (March 1977) 34-40.

Bassett, G. A., "What Is Communication and How Can I Do It Better?," COMMUNICATING EFFECTIVELY, New York: AMACON, Div. of American Management Associations (1975) 32-39.

Bennigson, L. A., "The Strategy of Running Temporary Projects," INNOVATION, No. 24 (September 1977) 32-40.

Cammann, C., and Nadler, D. A., "Fit Control Systems to Your Managerial Style," HARVARD BUSINESS REIVEW, *54*, No. 1 (January-February 1976) 65-72.

Chopra, A., "Motivation in Task-Oriented Groups," J. NURSING ADMINISTRATION, *3*, No. 1 (January-February 1973) 55-60.

Cleland, D. I., and King, W. R., SYSTEMS ANALYSIS AND PROJECT MANAGEMENT, 2nd ed., New York: McGraw-Hill Book Co. (1975) xviii + 398 pp.

Clough, R. H., and Sears, G. A., CONSTRUCTION PROJECT MANAGEMENT, 2nd ed., New York: Wiley-Interscience (1979) x + 341 pp.

Doyle, M., and Straus, D., HOW TO MAKE MEETINGS WORK: THE NEW INTERACTION METHOD, New York: Playboy Paperbacks, Div. of P.E.I. Books, Inc. (1977) x + 301 pp.

Dunsing, R. J., YOU AND I HAVE SIMPLY GOT TO STOP MEETING THIS WAY, New York: AMACON, Div. of American Management Associations (1977) 88 pp.

Estes, W. E., and Ruskin, A. M., "The Organization: Its Effect on Project Managers," PROCEEDINGS OF THE 1980 ASCE FALL CONVENTION, New York: American Society of Civil Engineers (October 1980).

Fiedler, F. E., and Chemers, M. M., LEADERSHIP AND EFFECTIVE MANAGEMENT, Glenview, Illinois: Scott, Foresman & Co. (1974) 166 pp.

Fisher, R., and Ury, W., GETTING TO YES: NEGOTIATING AGREEMENT WITHOUT GIVING IN, Boston, Massachusetts: Houghton Mifflin Company (1981) xiii + 165 pp.

Fogel, I. M., "Managing the Small and Medium-Size Project," CHEMICAL ENGINEERING, *78*, No. 28 (December 13, 1971) 107-114.

Gido, J., AN INTRODUCTION TO PROJECT PLANNING, Schenectady, New York: General Electric Co. (1974) x + 168 pp.

Hajek, V. G., MANAGEMENT OF ENGINEERING PROJECTS, New York: McGraw-Hill Book Co. (1977) xvi + 264 pp.

Hansen, G. L., "Eight Basic Tasks for Successful Project Management," MANAGE, *26*, No. 2 (March-April 1974) 12-15.

Hersey, P., and Blanchard, K. H., MANAGEMENT OF ORGANIZATION BEHAVIOR: UTILIZING HUMAN RESOURCES, Englewood Cliffs, New Jersey: Prentice-Hall, Inc. (1977) xvi + 360 pp.

Herzberg, F., "One More Time: How Do You Motivate Employees?," HARVARD BUSINESS REVIEW, *46*, No. 1 (January-February) 1968) 53-62.

Hines, W. W., III, "Increasing Team Effectiveness," TRAINING AND DEVELOPMENT J., *34*, No. 2 (February 1980) 78-82.

Jenett, E., "Guidelines for Successful Project Management," CHEMICAL ENGINEERING, *80*, No. 16 (July 9, 1973) 70-82.

Karrass, C. L., GIVE AND TAKE, New York: Thomas Y. Crowell, Publishers (1974) xv + 280 pp.

Karrass, C. L., THE NEGOTIATING GAME, New York: The World Publishing Company (1970) xii + 243 pp.

Kelley, A. J., and Murphy, D. C., "Project Management: Factors Leading to Success or Failure," PROCEEDING OF THE AIAA 10TH ANNUAL MEETING, New York: American Institute of Aeronautics and Astronautics (January 1974).

Kerzner, H., PROJECT MANAGEMENT: A SYSTEMS APPROACH TO PLANNING, SCHEDULING, AND CONTROLLING, New York: Van Nostrand Reinhold (1979) xii + 487 pp.

Ludwig, E. L., APPLIED PROJECT MANAGEMENT FOR THE PROCESS INDUSTRIES, Houston, Texas: Gulf Publishing Co. (1974) xi + 367 pp.

Maciariello, J. A., PROGRAM MANAGEMENT CONTROL SYSTEMS, New York: Wiley-Interscience (1978) xviii + 220 pp.

Maieli, V., "Sowing the Seeds of Project Cost Overruns," MANAGEMENT REVIEW, *61*, No. 8 (August 1972) 7-14.

Martin, C. C., PROJECT MANAGEMENT—HOW TO MAKE IT WORK, New York: AMACON, Div. of American Management Associations (1976) vi + 312 pp.

Martin, D. D., and Shell, R. L., WHAT EVERY ENGINEER SHOULD KNOW ABOUT HUMAN RESOURCES MANAGEMENT, New York: Marcel Dekker, Inc. (1980) 192 pp.

McClelland, D. C., and Burnham, D. H., "Power is the Great Motivator," HARVARD BUSINESS REVIEW, *54*, No. 2 (March-April 1976) 100-110.

Metzger, P. W., MANAGING A PROGRAMMING PROJECT, 2nd ed., Englewood Cliffs, New Jersey: Prentice-Hall, Inc. (1981) xi + 244 pp.

Miller, S. Wackman, D., Nunnally, E., and Saline, C., STRAIGHT TALK, New York: Rawson, Wade Publishers, Inc. (1981) xiv + 340 pp.

Mintz, H. K., "Memos That Get Things Moving," COMMUNICATING EFFECTIVELY, New York: AMACON, Div. of American Management Associations (1975) 75-82.

Myers, I. B., GIFTS DIFFERING, Palo Alto, California: Consulting Psychologists Press, Inc. (1980) 230 pp.

Myers, I. B., INTRODUCTION TO TYPE, Palo Alto, California: Consulting Psychologists Press, Inc. (1980) 18 pp.

Myers, I. B., MANUAL, MYERS-BRIGGS TYPE INDICATOR, Palo Alto, California: Consulting Psychologists Press, Inc. (1962) ii + 110 pp.

Nierenberg, G. I., CREATIVE BUSINESS NEGOTIATING, New York: Hawthorn Books, Inc. (1971) x + 182 pp.

Nierenberg, G. I., THE ART OF NEGOTIATING, New York: Cornerstone Library, Inc. (1980) 192 pp.

O'Brien, J. J., CPM IN CONSTRUCTION MANAGEMENT: SCHEDULING BY THE CRITICAL PATH METHOD, New York: McGraw-Hill Book Co. (1965) ix + 254 pp.

O'Brien, J. J., ed., SCHEDULING HANDBOOK, New York: McGraw-Hill Book Co. (1969) x + 605 pp.

Ostwald, P. F., COST ESTIMATING FOR ENGINEERING AND MANAGEMENT, Englewood Cliffs, New Jersey: Prentice-Hall, Inc. (1974) xiii + 493 pp.

Park, W. R., COST ENGINEERING ANALYSIS: A GUIDE TO THE ECONOMIC EVALUATION OF ENGINEERING PROJECTS, New York: Wiley-Interscience (1973) x + 308 pp.

Peterson, P., "Project Control Systems," DATAMATION, *25*, No. 7 (June 1979) 147-162.

Pilcher, R., APPRAISAL AND CONTROL OF PROJECT COSTS, London: McGraw-Hill Book Company (1973) xii + 324 pp.

Porter, E. H., and Maloney, S. E., MANUAL, STRENGTH DEPLOYMENT INVENTORY, Pacific Palisades, California: Personal Strengths Publishing, Inc. (1977) 31 pp.

Porter, E. H., "On the Development of Relationships Awareness Theory: A Personal Note," GROUP AND ORGANIZATIONAL STUDIES, *1*, No. 3 (September 1976) 302-309.

Project Management Institute, SURVEY OF CPM SCHEDULING SOFTWARE PACKAGES AND RELATED PROJECT CONTROL

PROGRAMS, Drexel Hill, Pennsylvania (January 1980) 129 pp.

Rosenau, M. D., Jr., SUCCESSFUL PROJECT MANAGMENT, Belmont, California: Lifetime Learning Publications (1981) xv + 266 pp.

Ruskin, A. M., "Monitoring and Contingency Allowances: Complementary Aspects of Project Control," PROJECT MANAGEMENT QUARTERLY, *12*, No. 4 (December 1981) 49-50.

Ruskin, A. M., and Lerner, R., "Forecasting Costs and Completion Dates for Defense Research and Development Contracts," IEEE TRANSACTIONS ON ENGINEERING MANAGEMENT, *EM-19*, No. 4 (November 1972) 128-133.

Silverman, M., PROJECT MANAGEMENT—A SHORT COURSE FOR PROFESSIONALS, New York: Wiley Professional Development Programs (1976) 300 pp.

Stewart, J. M., "Making Project Management Work," BUSINESS HORIZONS, *8*, No. 3 (Fall 1965) 54-68.

Stogdill, R. M., HANDBOOK OF LEADERSHIP, New York: The Free Press (1974) viii + 613 pp.

Stuckenbruck, L. C., ed., THE IMPLEMENTATION OF PROJECT MANAGEMENT: THE PROFESSIONAL'S HANDBOOK, Reading, Massachusetts: Addison-Wesley Publishing Co. (1981) xii + 254 pp.

Thamhaim, H. J., and Wilemon, D. L., "Leadership, Conflict, and Program Management Effectiveness," SLOAN MANAGEMENT REVIEW, *19*, No. 1 (Fall 1977) 69-89.

Webster, F. M., "Ways to Improve Performance on Projects," PROJECT MANAGEMENT QUARTERLY, *12*, No. 3 (September 1981) 21-26.

Wilemon, D. L., and Gemmill, G. R., "Interpersonal Power in Temporary Management Systems," J. MANAGEMENT STUDIES, *8*, No. 3 (October 1971) 315-328.

Youker, R., "Organization Alternatives for Project Managers," MANAGEMENT REVIEW, *66*, No. 11 (November 1977) 46-53.

Zeldman, M., KEEPING TECHNICAL PROJECTS ON TARGET, New York: AMACON, Div. of American Management Associations (1978) 44 pp.

Zemke, R., "Team Building: Helping People Learn to Work Together," TRAINING HRD, *15*, No. 2 (February 1978) 23-26.

Index